Artificial Intelligence: Intelligence that Matters

Innovative Perspectives & Essays

on

Artificial Intelligence

Dr. Bikram Lamba

"A powerful AI system tasked with ensuring your safety might imprison you at home. If you asked for happiness, it might hook you up to a life support and ceaselessly stimulate your brain's pleasure centers. If you don't provide the AI with a very big library of preferred behaviors or an ironclad means for it to deduce what behavior you prefer, you'll be stuck with whatever it comes up with. And since it's a highly complex system, you may never understand it well enough to make sure you've got it right."

— Barrat, James. Our Final Invention: Artificial Intelligence and the End of the Human Era

"If an AI possessed any one of these skills—social abilities, technological development, economic ability—at a superhuman level, it is quite likely that it would quickly come to dominate our world in one way or another. And as we've seen, if it ever developed these abilities to the human level, then it would likely soon develop them to a superhuman level. So we can assume that if even one of these skills gets programmed into a computer, then our world will come to be dominated by AIs or AI-empowered humans."

— Armstrong, Stuart. Smarter Than Us: The Rise of Machine Intelligence

Table of Contents

Preface

Artificial Intelligence is the hall-mark of new dawn. It is providing a new dimension to the Fourth Industrial revolution. What is it? It has come a long way since it was first created. Allen Newell, J. C. Shaw, and Herbert A. Simon pioneered the newly created artificial intelligence field with the Logic Theory Machine (1956), and the General Problem Solver in 1957. In 1958, John McCarthy and Marvin Minsky started the MIT Artificial Intelligence lab with $50,000. It is essentially Machine Learning that is a type of artificial intelligence (AI) that provides computers with the ability to learn without being explicitly programmed. Machine learning focuses on the development of computer programs that can change when exposed to new data.

I have in my book, endeavored to provide focus on diverse aspects of this attribute of human development that has immense societal, technological, cultural, and even civilizational variations and holds awesome possibilities for the future of economic development and civilizational and ethical alteration.

The essays are an attempt to elucidate and explain the different dimensions, as well as study those characteristics that are either given a go-by or just glossed over.

The essays also take into consideration the fears that many people entertain: fears about economy, unemployment and even the emergence of a situation that might make human being surplus.

The essays are explanatory and provocative; and mean to inform you, and make you think. At times, they are academic in nature, but the underlying purpose is to make you think and react.

Innisfil   Bikram Lamba
June 2017

On the Cusp of an AI Revolution

Over the last 30 years, consumers have reaped the benefits of dramatic technological advances. In many countries, most people now have in their pockets a personal computer more powerful than the mainframes of the 1980s. The Atari 800XL computer that my son played games on when he was in high school was powered by a microprocessor with 3,500 transistors; the computer running on my iPhone today has two billion transistors. Back then, a gigabyte of storage cost $100,000 and was the size of a refrigerator; today it's basically free and is measured in millimeters.

Even with these massive gains, we can expect still faster progress as the entire planet – people and things – becomes connected. Already, five billion people have access to a mobile device, and more than three billion people can access the Internet. In the coming years, 50 billion things – from light bulbs to refrigerators, roads, clothing, and more – will be connected to the Internet as well.

Every generation or so, emerging technologies converge, and something revolutionary occurs. For example, a maturing Internet, affordable bandwidth and file-compression, and Apple's iconic iPhone enabled companies such as Uber, Airbnb, YouTube, Facebook, and Twitter to redefine the mobile-customer experience.

Now we are on the cusp of another major convergence: big data, machine learning, and increased computing power will soon make artificial intelligence, (AI), ubiquitous.

AI follows Albert Einstein's dictum that genius renders simplicity from complexity. So, as the world itself becomes more complex, AI will become the defining technology of the twenty-first century just as the microprocessor was in the twentieth century.

Consumers already encounter AI on a daily basis. Google uses machine learning to autocomplete search queries and often accurately predicts what someone is looking for. Facebook and Amazon use predictive algorithms to make recommendations based on a user's reading or purchasing history. AI is the central component in self-driving cars – which can now avoid collisions and traffic congestion – and in game-playing systems like Google DeepMind's AlphaGo, a computer that beat South Korean Go master Lee Sedol in a five-game match earlier in 2016.

Given AI's wide applications, all companies today face an imperative to integrate it into their products and services; otherwise, they will not be able to compete with companies that are using data-collection networks to improve customer experiences and inform business decisions. The next generation of consumers will have grown up with digital technologies and will expect companies to anticipate their needs and provide instant, personalized responses to any query.

So far, AI has been too costly or complex for many businesses to make optimal use of it. It can be difficult to integrate into a business's existing operations, and historically it has required highly skilled data scientists. As a result, many businesses still make important decisions based on instinct instead of information.

This will change in the next few years, as AI becomes more pervasive, potentially making every company and every employee smarter, faster, and more productive. Machine learning algorithms can analyze billions of signals to route customer service calls automatically to the most appropriate agent or determine which customers are most likely to purchase a particular product.

And AI's applications extend beyond online retail: Brick-and-mortar stores still account for 90% of retail sales, according to the consultancy A.T. Kearney. Soon, when customers enter a physical store, they will be greeted by interactive chat-bots that can recommend products based on their shopping history, offer special discounts, and handle customer-service issues.

Advances in so-called "deep learning," a branch of AI modeled after the brain's neural network, could enable intelligent digital assistants to help plan vacations with the acumen of a human assistant, or determine consumer sentiments toward a particular brand, based on millions of signals from social networks and other data sources. In health care, deep-learning algorithms could help doctors identify cancer-cell types or intracranial abnormalities from anywhere in the world in real time.

To deploy AI effectively, companies will need to keep privacy and security in mind. Because AI is fueled by data, the more data the machine gains about an individual, the better it can predict their needs and act on their behalf. But, of course, that massive flow of personal data could be appropriated in ways that breach trust. Companies will have to be transparent about how they use people's personal data. AI can also detect and defend against digital security breaches, and will play a critical role in protecting user privacy and in building trust.

As in past periods of economic transformation, AI will unleash new levels of productivity, augment our personal and professional lives, and pose existential questions about the age-old relationship between man and machine. It will disrupt industries and dislocate workers as it automates more tasks. But just as the Internet did 20 years ago, AI will also improve existing jobs and spawn new ones. We should expect this and adapt accordingly by providing training for the jobs of tomorrow, as well as safety nets for those who fall behind.

AI is still a long way from surpassing human intelligence. It has been 60 years since John McCarthy, a computer scientist and nominal father of AI, first introduced the term during a conference at Dartmouth College, and computers have only recently been able to detect cats in YouTube videos or determine the best route to the airport.

We can count on technological innovation to continue at an even more rapid pace than in previous generations. AI will become like electrical current – invisible and augmenting almost every part of our lives. Thirty years from now, we will wonder how we ever got along without our seemingly telepathic digital assistants, just as today it's already hard to imagine going more than a few minutes without checking the 1980s' mainframe in one's pocket.

Resistance to New Technologies

Change is the law of nature, but the nature of human beings is to resist any change. Humans are averse to any change and accept change over time. Flux or change just disturbs the equanimity of mankind. It is a rare breed of revolutionaries that welcomes change as a matter of life and takes it up as an act of improvement. Microsoft founder Bill Gates doesn't understand why people are not concerned about artificial intelligence (AI), agreeing with Elon Musk that it could be one of our biggest existential threats, but Microsoft's research head Eric Horvitz disagrees with this view. Concern over the social and economic impacts of AI is one of the many controversies surrounding emerging technologies.

There are many reasons for this opposition to new technologies. Essentially, it is our sense of what it means to be human lies at the root of some of the skepticism about technological innovation. Given Schumpeter's comments on innovators and entrepreneurs – he once said that their work opened them up to "social ostracism and to physical prevention or to direct attack" – there could not have been a more suitable venue. Schumpeter wrote this comment in 1912, which is to say that we have a long history of resisting technological advances. And it's to history we must turn to understand why this is so.

Looking in the past for answers

The society tends to reject new technologies when they substitute for, rather than augment, our humanity. Our desire to humanize technology is captured in the humor of this Bradley's Bromide: "If computers get too powerful, we can organize them into a committee – that will do them in." We eagerly embrace them when they support our desire for inclusion, purpose, challenge, meaning and alignment with nature. We do so even when they are unwieldy, expensive, time-consuming to use, and constantly break down.

For example, the early days of the introduction of tractors in the United States were hardly the paragon of farm efficiency. Tractors offered little advantage over horses. Some opponents argued that their value could be marginally improved if they could reproduce themselves like horses.

What 'brick phones' teach us about new technologies

As technologies migrate across countries and continents, their societal implications also change. For example, when Motorola introduced cellphones in the United States in 1983, they were dismissed as toys for the rich. They cost $4,000 (today's equivalent of $10,000), weighed two pounds, stood at a foot tall, took 10 hours to charge, and delivered only 30 minutes of talk time. These metrics would have qualified them as a tool for updating one's Facebook status. They were the butt of jokes, dubbed "brick phones" because of their shape and weight.

The first model was called DyanTAC, standing for Dynamic Adaptive Total Area Coverage. Despite this aggressive and prospective name, the early models did little to augment our humanity, especially for young people. Adoption rates in the United States were glacial, putting it well behind Europe, Asia, and Africa.

When cellphones hit Africa, they were reinvented by engineers and diffused using novel business models created by entrepreneurs in Kenya, who pioneered mobile money transfer – called "transfer" instead of "banking" because banks wouldn't let the telecoms hold money.

Today cell phones are no longer just a communication tool. They are serving as banks, schools, clinics, and vehicles for spreading transparency and democracy. They augment our humanity in ways that could not have been anticipated in the early 1980s. They are also serving as a role model for improvements in other sectors such as off-grid electricity supply.

And now we have more than just cell phones. We live in exciting times where technological diversity and creativity offer limitless opportunities to expand the human potential for all, not just for certain exclusive sections of society.

When technologies 'give back'

Resistance to new technologies is heightened when the public perceives that the benefits of new technologies will only accrue to a small section of society, while the risks are likely to be widespread. Therefore, technologies promoted by large corporations often face stiff opposition from the public.

Similarly, new technologies face great opposition when the public perceives that the risks are likely to be felt in the short run and the benefits will only accrue eventually. So telling a skeptical public that new technologies will benefit future generations does not protect us from the wrath of current ones.

What is the way forward? The answer might lie in the much-abused phrase "social entrepreneurship". For many, this term is a euphemism for a charity or nongovernmental organization. But what is really needed is to bring the "social" back into "entrepreneurship". This means exploring new ways by which enterprises can be seen as contributing to the common good. The fact that enterprises use new technologies to enhance their competitiveness makes it difficult for the general public to separate technology from its uses – for better or for worse.

The fate of new technologies will continue to be determined by the balance of power in society. For nearly 400 years, Ottoman rulers opposed the printing of the Koran. Doing so would have undermined the role of religious leaders as sources of cultural codes. But when the printed word seemed to reinforce the power of the rulers they slowly went against previous fatwas banning the printing of the Koran.

The acceptable aspect of new technologies is dependent on whether they reinforce rather than undermine incumbent practices. The dilemma facing modern society is whether reinforcing existing practices undermines society.

New technologies are essential to fostering economic growth, meeting human needs, and protecting the environment. New clean energy technologies such as solar photovoltaic cells and wind turbines, for example, are critical to reducing carbon dioxide emission and addressing the challenges of climate change. But their adoption is often held back by the incumbent industries and vested interests. The dilemma is that in many cases clinging to the old may, in fact, be in conflict with our humanity, especially in regard to our search for affinity with nature. As the American composer John Cage aptly put it: "I can't understand why people are frightened of new ideas. I'm frightened of the old ones."

Fears about Technology are Nothing New

We live, we are so often told, in an information age. It is an era obsessed with space, time, and speed, in which social media inculcates virtual lives that run parallel to our "real" lives and in which communication technologies collapse distances around the globe. Many of us struggle with the bombardment of information we receive and experience anxiety because of new media, which we feel threaten our relationships and "usual" modes of human interaction.

Though the technologies may change, these fears have a very long history: more than a century ago our forebears had the same concerns. Literary, medical, and cultural responses in the Victorian age to the perceived problems of stress and overwork anticipated many of the preoccupations of our own era to an extent that is perhaps surprising. This parallel is well illustrated by the 1906 cartoon from Punch, a satirical British weekly magazine: the caption reads:

"These two figures are not communicating with one another. The lady receives an amatory message, and the gentleman some racing results." The development of the "wireless telegraph" is portrayed as an overwhelmingly isolating technology.

Replace these strange contraptions with smartphones, and we are reminded of numerous contemporary complaints regarding the stunted social and emotional development of young people, who no longer hang out in person, but in virtual environments, often at great physical distance. Different technology, same statement. And it's underpinned by the same anxiety that "real" human interaction is increasingly under threat from technological innovations that we have, consciously or unconsciously, assimilated into daily life. By using such devices, so the popular paranoia would have it, we are somehow damaging ourselves.

Cacophony of voices

The 19th century witnessed the rapid expansion of the printing industry. New techniques and mass publishing formats gave rise to a far more pervasive periodical press, reaching a wider readership than ever before. Many celebrated the possibility of instant news and greater communication. But concerns were raised about the overwhelmed middle-class reader who, it was thought, lacked the discernment to critically judge the new mass of information, and so read everything in a superficial, erratic manner.

The philosopher and essayist Thomas Carlyle, for example, lamented the new lack of direct contact with society and nature caused by the intervention of machinery in every aspect of life. Print publications were fast becoming the principal medium of public debate and influence, and they were shaping and, in Carlyle's view, distorting human learning and communications.

The philosopher and economist John Stuart Mill heartily agreed, expressing his fears in an essay entitled "Civilisation". He thought that the cacophony of voices supposedly overwhelming the public was creating a state of society where any voice, not pitched in an exaggerated key, is lost in the hubbub. Success in so crowded a field depends not upon what a person is, but upon what he seems: mere marketable qualities become the object instead of substantial ones, and a man's capital and labour are expended less in doing anything than in persuading other people that he has done it. Our own age has seen this evil brought to its consummation.

Individual authors and writers were becoming disempowered, lost in a glutted marketplace of ideas, opinions, adverts, and quacks.

Old complaints

The parallels with the concerns of our own society are striking. Arguments along not at all dissimilar lines have been advanced against contemporary means of acquiring information, such as Twitter, Facebook, and our constant access to the internet in general.

In his 2008 article, "Is Google Making Us Stupid?", journalist Nicolas Carr speculated that "we may well be in the midst of a sea change in the way we read and think". Reading online, he posits, discourages long and thoughtful immersion in texts in favour of a form of skipping, scanning, and digressing via hyperlinks that will ultimately diminish our capacity for concentration and contemplation.

Writers, too, have shared Carr's anxieties. Philip Roth and Will Self, for example, have both prophesied these trends as contributing to the death of the novel, arguing that people are increasingly unused to and ill-equipped to engage with its characteristically long, linear form.

Of course, all old technologies were once new. People were at one point genuinely concerned about things we take for granted as perfectly harmless now.

In the later decades of the 19th century it was thought that the telephone would induce deafness and that sulphurous vapours were asphyxiating passengers on the London Underground. These then-new advancements were replacing older still technologies that had themselves occasioned similar anxieties on their introduction. Plato, as his oral culture began to transition to a literary one, was gravely worried that writing itself would erode the memory.

While we cannot draw too strict a line of comparison

between 19th-century attitudes to such technologies as the telegraph, train, telephone, and newspaper and our own responses as a culture to the advent of the internet and the mobile phone, there are parallels that almost argue against the Luddite position. As dramatically as technology changes, we, at least in the way we regard it, remain surprisingly unchanged.

AI Leads to Business Change

Artificial Intelligence (AI) may be the single most disruptive technology the world has seen since the Industrial Revolution. Granted, there is a lot of hype out there on AI, along with doomsday headlines and scary movies. But the reality is that it will positively and materially change how we engage with the world around us. It's going to improve not only how business is done, but the kind of work we do – and unleash new levels of creativity and ingenuity.

In fact, research from Accenture estimates that artificial intelligence could double annual economic growth rates of many developed countries by 2035, transforming work and fostering a new relationship between humans and machines. The report projects that AI technologies in business will boost labor productivity by up to 40 percent. Rather than undermining people, we believe AI will reinforce their role in driving business growth. As AI matures, it will potentially serve as a powerful antidote for the stagnant productivity and shortages in skilled labor of recent decades.

While it is early, we are already seeing AI's impact. Combined with cloud, sophisticated analytics and other technologies, it is starting to change how work is done by both people and computers. It's also changing how organizations interact with consumers: sometimes in startling ways!

AI is flourishing now because of the rise of ubiquitous computing, low-cost cloud services, near unlimited inexpensive storage, new algorithms, and other related technology innovations. Cloud computing along with advances in Graphical Processing Units (GPU's) has provided necessary computational power. AI algorithms and architectures have progressed rapidly, often enabled by open source software.

Artificial Intelligence is not just one technology, but rather a variety of different sorts of software that can be applied in numerous ways for different applications.

But equally important is a vast increase in the availability of data. AI does not think for itself. Its insights are possible because the software gets fed information, and the more information it gets, the more insight it can produce. Over the last decade, crowdsourced data in particular has proliferated on internet and social media. People in their daily lives upload massive quantities of images, videos, social media comments, and chat dialogues. All that creates labelled data that is available for machines to use in what's called machine learning.

While many believe that AI will supplant humans, we think it will instead mostly enable people to do more exceptional work. Certainly, AI will cause displacement of jobs, but it may also significantly boost the productivity of labor. Innovative AI technologies will enable people to make more efficient use of their time and do what humans do best – create, imagine, and innovate new things.

With technology, overall and AI in particular, the key ingredient for success and creating value is taking a "people first" approach. But to make this transition means both companies and governments must acknowledge the challenges and change how they behave. They must be thoroughly prepared—intellectually, technologically, politically, ethically, and socially. Governments and businesses will need to take several steps, many of which are not easy:

Prepare the next generation:

Re-evaluate the type of knowledge and skills required for the future, and address the need for education and training. AI presents the opportunity to prepare an entirely new class of skilled and trained workers that does not exist today. This training should be targeted to help those who are disproportionately affected by the coming changes in employment and incomes.

Advocate for and develop a code of ethics for AI.

Ethical debates, challenging as they will be, should be supplemented by tangible standards and best practices in the development and use of intelligent machines.

Encourage AI-powered regulation.

Update old laws and use AI itself to create adaptive, self-improving new ones to help close the gap between the pace of technological change and the pace of regulatory response. This will require government to think and act in new ways appropriate to the new landscape; and means more technologically-trained people must play an active role in government.

Work to integrate human intelligence with machine intelligence.

Businesses must begin reimagining business processes, and reconstructing work to take advantage of the respective strengths of people and machines.

The market demand and opportunity for AI is expanding rapidly, with analyst firm IDC predicting that the worldwide content analytics, discovery and cognitive systems software market will grow from $4.5 billion in 2014 to $9.2 billion in 2019. In fact, Accenture's Technology Vision 2016—research that gathers input from more than 3,100 global business and IT executives—found that 70% of them are making significantly more investments in AI-related technologies than two years ago, with 55% planning to use machine learning and embedded artificial intelligence. Equity financings for AI companies have risen from $282 million in 2011 to $2.4 billion in 2015, or 746%, according to researchers at CB Insights. AI patents are being granted at a rate five times greater than 10 years ago. AI start-ups in the US alone have increased 20-fold in just 4 years.

A major Italian government agency offers a good example of how AI can dovetail with the work people do and enable them to be more effective. Employees there were spending the majority of their time attending to routine customer queries. The agency worked with Accenture to automate the process with AI. An intelligent Virtual Agent application now handles real-time voice calls and webchat interactions, using a combination of cognitive-semantic analysis and machine-learning algorithms. After just three months, the Virtual Agent application has already successfully served more than 70,000 users. Employees can now take on more demanding and rewarding tasks, which can positively impact their engagement.

AI is also positively impacting how governments operate. The Singapore government's Safe City program uses the latest in video analytics and image recognition to assist in public safety. It increases security, delivers services more effectively and makes more efficient use of city resources.

The Accenture Institute for High Performance and Accenture Technology, in collaboration with Frontier Economics, modeled the impact of artificial intelligence on 12 developed economies that together generate more than 50 percent of the world's economic output. The research compared the size of each country's economy in

2035 under a baseline scenario, in which economic growth continues under current conditions, with an AI scenario, in which the impact of AI has been absorbed into the economy.

AI was found to yield the highest economic benefits for the United States, increasing annual growth from 2.6 percent to 4.6 percent by 2035, translating to an additional $8.3 trillion in gross value added (GVA). In the United Kingdom, AI could add an additional $814 billion to the economy in the same period. Japan has the potential to more than triple its annual rate of GVA growth by 2035, and Finland, Sweden, the Netherlands, Germany and Austria could see their growth rates double.

AI can empower people to create, imagine and innovate at entirely new levels to drive growth and productivity. Far from simply eliminating repetitive tasks, AI should put people at the center, augmenting the workforce by applying the capabilities of machines so people can focus on higher-value analysis, decision-making and innovation.

Curb the Artificial Intelligence Arms Race

The machines rise, subjugating humanity. It's a science fiction trope that's almost as old as machines themselves. The doomsday scenarios spun around this theme are so outlandish: like *The Matrix*, in which human-created artificial intelligence plugs humans into a simulated reality to harvest energy from their bodies. It's difficult to visualize them as serious threats.

Meanwhile, artificially intelligent systems continue to develop apace. Self-driving cars are beginning to share our roads; pocket-sized devices respond to our queries and manage our schedules in real-time; algorithms beat us at Go; robots become better at getting up when they fall over. It's obvious how developing these technologies will benefit humanity. But then? Don't all the dystopian sci-fi stories start out this way?

Any discussion about the dystopian potential of AI risks gravitating towards one of two extremes. One is overly credulous scare-mongering. Of course, Siri isn't about to transmogrify into murderous HAL from *2001: A Space Odyssey*. But the other extreme, complacency, is equally dangerous. We don't need to think about these issues, because humanity-threatening AI is decades or more away.

It is true that the artificial "superintelligence" beloved of sci-fi may be many decades in the future, if it is possible at all. However, a recent survey of leading AI researchers by TechEmergence found a wide variety of concerns about the security dangers of AI in a much more realistic, 20-year timeframe; for example, it can cause financial system meltdown as algorithms interact unexpectedly, and the potential for AI to help malicious actors optimize biotechnological weapons.

These examples show how, alongside technological progress on many fronts, the Fourth Industrial Revolution is promising a rapid and massive democratization of the capacity to wreak havoc on a very large scale. On the dark side of the "deep web", where information is hidden from search engines, destructive tools across a range of emerging technologies, from 3D-printed weapons to fissile material and equipment for genetic engineering in home laboratories, already exist for sale. In each case, AI exacerbates the potential for harm.

Consider another possibility mentioned in the TechEmergence survey. If we combine a gun, a quadrocopter drone, a high-resolution camera, and a facial recognition algorithm that wouldn't need to be much more advanced than the current best in class, we could in theory make a machine we can program to fly over crowds, seeking particular faces and assassinating targets on sight.

Such a device would require no superintelligence. It is conceivable using current, "narrow" AI that cannot yet make the kind of creative leaps of understanding across distinct domains that humans can. When "artificial general intelligence", or AGI, is developed eventually, as seems likely, it will significantly increase both the potential benefits of AI and, in the words of Jeff Goddell, its security risks, "forcing a new kind of accounting with the technological genie".

But not enough thinking is being done about the weaponizable potential of AI. As Wendell Wallach puts it: "Navigating the future of technological possibilities is a hazardous venture". He further says, "It begins with learning to ask the right questions—questions that reveal the pitfalls of inaction, and more importantly, the passageways available for plotting a course to a safe harbor."

Non-proliferation challenges

Prominent scholars including Stuart Russell have issued a call for action to avoid "potential pitfalls" in the development of AI that has been backed by leading technologists including Elon Musk, Demis Hassabis, Steve Wozniak and Bill Gates. One high-profile pitfall could be "lethal autonomous weapons systems" (LAWS) – or, more colloquially, "killer robots". Technological advances in robotics and the digital transformation of security has already changed the fundamental paradigm of warfare. According to Christopher Coker "21st-century technology is changing our understanding of war in deeply disturbing ways." Fully developed LAWS will likely transform modern warfare as dramatically as gunpowder and nuclear arms.

The U.N. Human Rights Council has called for a moratorium on the further development of LAWS, while other activist groups and campaigns have advocated for a full ban, drawing an analogy with chemical and biological weapons, which the international community considers beyond the pale. For the third year in a row, the United Nations Member States met to debate the call for a ban, and how to ensure that any further development of LAWS stays within international humanitarian law. However, when ground-breaking weapons technology is no longer confined to a few large militaries, non-proliferation efforts become much more difficult.

The debate is complicated by the fact that definitions remain mired in confusion. Platforms, such as drones, are commonly confused with weapons that can be loaded on platforms. The idea of systems being asked to execute narrowly defined tasks, such as to identify and eliminate armored vehicles moving in a specific geographical area, is not always distinguished from the idea of systems being given discretionary scope to interpret more general missions, such as "win the war".

Some argue to always keep a "human in the loop" to exercise "meaningful human control". This would preclude fully autonomous systems, which might view the "man" and the "law" as nothing but a nuisance to the system's ability to deliver its task. However, limiting systems to semi-autonomy or "sliding autonomy" – in which the "man" is brought in only in unusual circumstances – also has flaws.

Experience from testing self-driving cars suggests that humans struggle to stay alert and lose situational awareness when supervising a system that usually runs in automated mode. *Deepmind*, a company acquired by Google in 2014, has recently published a paper together with the Future of Humanity Institute at Oxford, where they describe an AI "off-switch" or a "big red button". The paper outlines a "framework" to allow a "human operator" to safely interrupt an AI. According to one of the authors: "our framework allows the human supervisor to temporarily take control of the agent and make it believe it/chooses/to shut down itself". However, the paper also clearly states that "it is unclear if all algorithms can be easily made safely interruptible". Add to that it "is hard to predict when human will need to start pressing a "big red button" on self-learning machines".

An additional concern is that any weapons system with a degree of autonomy could be spoofed and the programmed objectives corrupted remotely by a purpose-engineered virus like the Stuxnet worm.

It may already be too late to arrest the development of LAWS: Peter Singer and August Cole, the authors of Ghost Fleet, say "the ship may already have sailed – without the need of a crew". Depending on the definition of autonomy, an argument can be made that such systems are already in use – Israel's Harpy drone being the clearest-cut case.

There is little current support from governments for a full ban on LAWS. One reason is that the technologies needed to develop more advanced LAWS are likely to become widely available in time – and if it would be impractical to prevent a terrorist group like ISIS from developing killer robots, then states may want to ensure they understand the technology themselves. Another reason is cost-effectiveness: humans often make up most of a defense budget, and LAWS – especially "swarms", in which many small robots can perform tasks simultaneously – could potentially cut costs drastically.

Others even make a moral case in favour of state actors developing killer robots. They could reduce the number of soldiers being killed, or returning traumatized from battle. And suppose an algorithm is developed that performs better than the most highly-trained soldiers at coolly making snap decisions in the heat of battle, distinguishing civilians from combatants, and opening fire only on the latter: would humanitarian considerations not oblige military leaders to take the error-prone humans out of the equation? In the words of Cumming, How and Williams, "partnering human and computer abilities, could greatly enhance planning tasks in a chaotic environment."

The counterpoint is that politicians might be more ready to start wars when they are sending robots than humans into battle – and the technology, once developed, is likely to be used, sooner or later, by those with scant regard for humanitarianism.

As artificial intelligence becomes more capable, similar questions will occur in more and more contexts, many of which are difficult to even imagine: How might the new capability conceivably be weaponized? Is it desirable? If not, how to control or – more likely – monitor its development? What norms about its use could and should be established? And how could any restrictions be enforced? It is already difficult to enforce restrictions on developing physical weapons – and the challenge becomes even starker when the "weapon" is software. Rod Beckstrom coins this as the "Conundrum of Artificial Intelligence".

Unfortunately, there is a growing gap between those developing AI and those who should be party to such a conversation. Public sector decision-makers typically have little understanding of the complexity of the technological possibilities being created in myriad start-ups around the world. Meanwhile the technologists themselves often struggle to internalize the "dark side" of technologies they view as life-enhancing, and the consequent need to govern against misuse.

Dual-use innovations

For obvious reasons, militaries do not reveal all their work on weaponizing artificial intelligence. However, Russia recently unveiled their "Iron Man" humanoid military robot, aiming to minimize the risk to soldiers in dangerous situations. The US and Chinese militaries, among others, are also investing heavily in AI and robotics. The US "third offset" strategy explicitly aims to keep it ahead in the technology game.

The geopolitical dimension to the third offset strategy

indicates an incipient AI arms race. As US Deputy Defense Secretary Work put it: "our adversaries are pursuing enhanced human operation and it scares the crap out of us, frankly".

An AI arms race would be unlikely to be as stable as the Cold War stand-off involving mutually-assured destruction. A common concern among AI researchers in the recent TechEmergence survey was the difficulty of predicting what happens when artificial intelligences engage with each other.

In contrast to the Cold War paradigm of military-sponsored cutting-edge research eventually spawning private sector applications, militaries are not necessarily at the cutting edge. Potentially weaponizable, "dual use" AI is increasingly being developed first in the private sector. For example, quadrocopter development is driven by commercial aims such as package deliveries. Facial recognition algorithms have a broad array of private sector as well as public security applications, such as recognizing when valued customers enter a store. According to Mary Cummings, the prominent robotics professor and former fighter pilot, "I guarantee you, Google and Amazon will soon have much more surveillance capability with drones than the military". She asks, "What happens when our governments are looking to corporations to provide them with the latest defense technology?"

The robotics race right now is causing a massive brain drain from militaries into the commercial world. The most talented minds are now being drawn towards the rewards offered in the private sector. Google's AI budget would be the envy of any military, and it can leverage its commercial activities to further research – for example, launching a photo storage service which will help refine its facial recognition software.

The significance of the private sector taking the lead is enormous: when technologies can be bought off-the-shelf, AI is potentially weaponizable by any non-state actor. Sooner or later, it will become trivially easy for organized criminal gangs or terrorist groups to construct devices such as assassination drones. Indeed, it is likely that given time, any AI capability that can be weaponized will be weaponized.

As AI develops, early attempts to weaponize it are likely to be buggy and prone to misfiring. But another implication of the brain drain from the military to private sector is a reduction in capacity to test and verify the effectiveness of technology to a degree that would instill confidence in battle situations. Legitimate actors may not want to send a technology that is considered only 80 percent ready into the battlefield.

Rogue actors, though, are unlikely to care about compliance or a bit of collateral damage. A terrorist organisation such as ISIS might be only too willing to use an 80 percent-ready AI weapon, with devastating results.

Towards superintelligence

Looking further to the future, the question that most fascinates sci-fi storytellers is what happens when an artificial general intelligence works out how to improve itself. Even today, the deep neural networks running narrow AI applications cannot be fully understood by the engineers who program them. A "superintelligence" could act in ways that defy human comprehension. Part of the challenge is verification. According to Alan Winfield "current verification approaches typically assume that the system being verified will never change its behaviour, but a system that learns does – by definition – change its behaviour, so any verification is likely to be rendered invalid after the system has learned." Verification is further complicated by what he coins as the "black box problem" describing the Artificial Neural Networks (ANNs) or rather the large sets of data that underpin all algorithms to make decisions and learn. The ANN is "trained" using these sets of data – but exactly how decisions are being made is not clear.

Dystopian stories like *The Matrix* envisage such superintelligent machines developing their own goals. But perhaps the more likely threat comes from another kind of story altogether. Advanced AGI may doom humanity not because it pursues its own goals, but because we fail to foresee some implication in the goals we set for it.

Scholars in the machine ethics community are increasingly thinking through these kinds of fundamental questions pertaining to AI. How do we instill human values in an artificial general intelligence, to forestall misunderstandings about what we want and curtail our own biases? What are human values, anyway? As Christopher Coker puts it in his book 'Future War': Will machines gradually "come to be seen not as replacements for human beings, but as extensions of our own humanity"?

Such discussions may still be academic rather than urgent concerns for policy-makers. But, as Stephen Hawking, Max Tegmark, Stuart Russell and Frank Wilczek stated in 2014, failing to take them seriously could be "potentially our worst mistake ever". It is now possible to foresee a continuum of AI development, from current narrow AI to possible superintelligence – and a structure is urgently needed to address the current security risks and keep abreast of them as AI develops.

The way forward

There is a need for a new, global platform to monitor, consider, and make recommendations about the implications of emerging technologies in general, and AI more specifically, for international security. Such a platform would have two imperatives.

The first is to build a multi-stakeholder platform to involve both the private and public sectors and decisionmakers in the dialogue. The purpose is to enable and encourage greater transparency about the capabilities of new inventions, even when weaponization could not be further from an innovator's mind.

The second is to find ways of moving beyond traditional, intergovernmental rule-making approaches to more creative regulation of new technologies. This will involve countries fundamentally rethinking their positions and expectations about non-proliferation efforts and disarmament processes, and considering practical measures to strengthen global, regional, and national norms.

Countries will always view any discussion on proliferation through a lens of their national security interests. But increasingly the security of all nations is interconnected. As technological progress democratizes the ability to inflict large-scale damage far beyond the historically important handful of major state actors, time-honored tools to prevent escalation of disputes – treaties, conventions, international organizations, game-theoretic concepts of deterrence – become less and less relevant.

Many AI applications have life-enhancing potential, so holding back its development is undesirable and possibly unworkable. This speaks for the need demanding a more connected and coordinated multistakeholder effort to create norms, protocols, and mechanisms for the oversight and governance of AI.

Gary Marchant and Wendell Wallach argue that emerging technologies, including AI, are better overseen by soft governance – industry initiatives, laboratory standards, testing and certification regimes, insurance policies – than hard governance, such as laws and regulations. This is an argument that holds promise in the overall non-proliferation discussion. Legal and regulatory regimes are typically slow-moving, while technological change is rapid; national, while innovation crosses borders; and stove-piped, while the biggest dangers often occur at the intersections of technologies.

Diplomats in the disarmament space have likewise argued that there is limited value in over-institutionalizing discussions on non-proliferation – this area of policy making needs to be agile by design. However, soft governance mechanisms are difficult to enforce, so it will still be necessary to put hard laws and regulatory bodies in place to forestall serious harms. In addition to the UN process referred to above, several joint and creative track-II multi-stakeholder governance initiatives for AI have begun. Other track-I processes are generally still lagging behind with a few exceptions. The World Economic Forum, as stated by Professor Schwab, its founder, will continue to use its platform to encourage dialogue and bring stakeholders together.

A new approach to the oversight and governance of AI would map the interests of relevant stakeholders as well as existing efforts to develop a shared concept on mitigating the security implications of AI. It would also enable strategies that reach beyond "headline technologies" such as killer robots, as well as look at the potentially destabilizing security effects of advanced AI capabilities upon unemployment and inequality. It would identify champions for a spirit of collaboration. It would work to debunk myths about AI, and identify gaps and blind spots. It would build a repository of knowledge and practices. It would further public and policy literacy on AI-related issues.

Transferable lessons from other processes and initiatives should be explored in greater depth where relevant. The 1990s Chemical Weapons Convention regarding non-proliferation space faced analogous issues of aligning business integrity with the need to test, verify and create a system for self-declaration of potentially relevant breakthroughs. Emerging multi-stakeholder regimes around the governance of cyberspace and climate change also offer insights, as do current discussions on relevant issues in driverless cars and aviation industry automation.

Above all, there is a need to recognize that humanity stands at an inflection point, with innovations in AI outpacing evolution in norms, protocols and governance mechanisms. A new and revived non-proliferation debate and architecture are needed to nurture holistic understanding of human relations with machines and automated systems, and influence the future trajectories and applications of AI and emerging technologies in general – making sure the outlandish, dystopian futures remain firmly in the realm of fiction.

Part Animal, Part Machine Robots

Think of a traditional robot and you probably imagine something made from metal and plastic. Such "nuts-and-bolts" robots are made of hard materials. As robots take on more roles beyond the lab, such rigid systems can present safety risks to the people they interact with. For example, if an industrial robot swings into a person, there is the risk of bruises or bone damage.

Researchers are increasingly looking for solutions to make robots softer or more compliant – less like rigid machines, more like animals. With traditional actuators – such as motors – this can mean using air muscles or adding springs in parallel with motors. For example, on a Whegs robot, having a spring between a motor and the wheel leg (Wheg) means that if the robot runs into something (like a person), the spring absorbs some of the energy so the person isn't hurt. The bumper on a Roomba vacuuming robot is another example; it's spring-loaded so the Roomba doesn't damage the things it bumps into.

But there's a growing area of research that's taking a different approach. By combining robotics with tissue engineering, we're starting to build robots powered by living muscle tissue or cells. These devices can be stimulated electrically or with light to make the cells contract to bend their skeletons, causing the robot to swim or crawl. The resulting biobots can move around and are soft like animals. They're safer around people and typically less harmful to the environment they work in than a traditional robot might be. And since, like animals, they need nutrients to power their muscles, not batteries, biohybrid robots tend to be lighter too.

Building a biobot

Researchers fabricate biobots by growing living cells, usually from heart or skeletal muscle of rats or chickens, on scaffolds that are nontoxic to the cells. If the substrate is a polymer, the device created is a biohybrid robot – a hybrid between natural and human-made materials.

If you just place cells on a molded skeleton without any guidance, they wind up in random orientations. That means when researchers apply electricity to make them move, the cells' contraction forces will be applied in all directions, making the device inefficient at best.

So to better harness the cells' power, researchers turn to micropatterning. We stamp or print microscale lines on the skeleton made of substances that the cells prefer to attach to. These lines guide the cells so that as they grow, they align along the printed pattern. With the cells all lined up, researchers can direct how their contraction force is applied to the substrate. So rather than just a mess of firing cells, they can all work in unison to move a leg or fin of the device.

Biohybrid robots inspired by animals

Beyond a wide array of biohybrid robots, researchers have even created some completely organic robots using natural materials, like the collagen in skin rather than polymers, for the body of the device. Some can crawl or swim when stimulated by an electric field. Some take inspiration from medical tissue engineering techniques and use long rectangular arms(or cantilevers) to pull themselves forward.

Others have taken their cues from nature, creating biologically inspired biohybrids. For example, a group led by researchers at California Institute of Technology

31

developed a biohybrid robot inspired by jellyfish. This device, which they call a medusoid, has arms arranged in a circle. Each arm is micropatterned with protein lines so that cells grow in patterns similar to the muscles in a living jellyfish. When the cells contract, the arms bend inwards, propelling the biohybrid robot forward in nutrient-rich liquid.

More recently, researchers have demonstrated how to steer their biohybrid creations. A group at Harvard used genetically modified heart cells to make a biologically inspired manta ray-shaped robot swim. The heart cells were altered to contract in response to specific frequencies of light – one side of the ray had cells that would respond to one frequency, the other side's cells responded to another.

When the researchers shone light on the front of the robot, the cells there contracted and sent electrical signals to the cells further along the manta ray's body. The contraction would propagate down the robot's body, moving the device forward. The researchers could make the robot turn to the right or left by varying the frequency of the light they used. If they shone more light of the frequency the cells on one side would respond to, the contractions on that side of the manta ray would be stronger, allowing the researchers to steer the robot's movement.

Toughening up the biobots

While exciting developments have been made in the field of biohybrid robotics, there's still significant work to be done to get the devices out of the lab. Devices currently have limited lifespans and low force outputs, limiting their speed and ability to complete tasks. Robots made from mammalian or avian cells are very picky about their environmental conditions. For example, the ambient temperature must be near biological body temperature and the cells require regular feeding with nutrient-rich liquid. One possible remedy is to package the devices so that the muscle is protected from the external environment and constantly bathed in nutrients.

Another option is to use more robust cells as actuators. At Case Western Reserve University, they've recently begun to investigate the possibility by turning to the hardy marine sea slug Aplysia Californica. Since A. Californica lives in the intertidal region, it can experience big changes in temperature and environmental salinity over the course of a day. When the tide goes out, the sea slugs can get trapped in tide pools. As the sun beats down, water can evaporate and the temperature will rise. Conversely in the event of rain, the saltiness of the surrounding water can decrease. When the tide eventually comes in, the sea slugs are freed from the tidal pools. Sea slugs have evolved very hardy cells to endure this changeable habitat.

The researchers have been able to use Aplysia tissue to actuate a biohybrid robot, suggesting that they can manufacture tougher biobots using these resilient tissues. The devices are large enough to carry a small payload – approximately 1.5 inches long and one inch wide.

A further challenge in developing biobots is that currently the devices lack any sort of on-board control system. Instead, engineers control them via external electrical fields or light. In order to develop completely autonomous biohybrid devices, they will need controllers that interface directly with the muscle and provide sensory inputs to the biohybrid robot itself. One possibility is to use neurons or clusters of neurons called ganglia as organic controllers.

This sea slug has been a model system for neurobiology research for decades. A great deal is already known about the relationships between its neural system and its muscles – opening the possibility that they could use its neurons as organic controllers that could tell the robot which way to move and help it perform tasks, such as finding toxins or following a light.

While the field is still in its infancy, researchers envision many intriguing applications for biohybrid robots. For example, the tiny devices using slug tissue could be released as swarms into water supplies or the ocean to seek out toxins or leaking pipes. Due to the biocompatibility of the devices, if they break down or are eaten by wildlife these environmental sensors theoretically wouldn't pose the same threat to the environment traditional nuts-and-bolts robots would.

One day, devices could be fabricated from human cells and used for medical applications. Biobots could provide targeted drug delivery, clean up clots or serve as compliant actuatable stents. By using organic substrates rather than polymers, such stents could be used to strengthen weak blood vessels to prevent aneurysms – and over time the device would be remodeled and integrated into the body. Beyond the small-scale biohybrid robots currently being developed, ongoing research in tissue engineering, such as attempts to grow vascular systems, may open the possibility of growing large-scale robots actuated by muscle.

A Cyber War Scenario

The internet has brought us many great things but it has made us more vulnerable. Imagine you woke up to discover a massive cyber-attack on your country. All government data has been destroyed: taking out healthcare records, birth certificates, social care records and so much more. The transport system isn't working; traffic lights are blank; immigration is in chaos and all tax records have disappeared. The internet has been reduced to an error message and daily life as you know it has halted.

This might sound fanciful but don't be so sure. When countries declare war on one another in future, this sort of disaster might be the opportunity the enemy is looking for. The internet has brought us many great things but it has made us more vulnerable. Protecting against such futuristic violence is one of the key challenges of the 21st century.

Strategists know that the most fragile part of internet infrastructure is the energy supply. The starting point in serious cyber warfare may well be to trip the power stations which power the data centers involved with the core routing elements of the network.

Back-up generators and uninterruptible power supplies might offer protection, but they don't always work and can potentially be hacked. In any case, backup power is usually designed to shut off after a few hours. That is enough time to correct a normal fault, but cyber-attacks might require backup for days or even weeks.

A major outage would cause large-scale economic damage and civil unrest throughout a country. In a war situation, this could be enough to bring about defeat. The American system is not well enough protected to avoid this.

Denial of service

An attack on the national grid could involve what is called a distributed-denial-of-service(DDoS) attack. These use multiple computers to flood a system with information from many sources at the same time. This could make it easier for hackers to neutralize the backup power and tripping the system.

DDoS attacks are also a major threat in their own right. They could overload the main network gateways of a country and cause major outages. Such attacks are commonplace against the private sector, particularly finance companies. Akamai Technologies, which controls 30% of internet traffic, are worrying about this kind of attack and becoming their ever more sophisticated.

Akamai recently monitored a sustained attack against a media outlet of 363 gigabits per second (Gbps) – a scale which few companies, let alone a nation, could cope with for long. There is a shocking 111% increase in DDoS attacks per year, almost half of them over 10 Gbps in scale – much more powerful than previously. The top sources are Vietnam, Brazil and Colombia.

Most DDoS attacks swamp an internal network with traffic via the DNS and NTP servers that provide most core services within the network. Without DNS, the internet wouldn't work; but it is weak from a security point of view. Specialists have been trying to come up with a solution, but building security into these servers to recognize DDoS attacks appears to mean re-engineering the entire internet.

How to react

If a country's grid were taken down by an attack for any length of time, the ensuing chaos would potentially be enough to win a war outright. If instead, its online infrastructure were substantially compromised by a DDoS attack, the response would probably go like this:

Phase one - Takeover of network: The country's security operations centre would need to take control of internet traffic to stop its citizens from crashing the internal infrastructure. This was seen in recent failed Turkish coup where YouTube and social media went completely offline inside the country.

Phase two - Analysis of attack: Security analysts would be trying to figure out how to cope with the attack without affecting the internal operation of the network.

Phase three - Observation and large-scale control: the authorities would be faced with countless alerts about system crashes and problems. The challenge would be to ensure only key alerts reached the analysts trying to overcome the problems before the infrastructure collapsed. A key focus would be ensuring military, transport, energy, health, and law enforcement systems along with financial systems, were given the highest priority.

Phase four - Observation and fine control: by this stage there would be some stability and the attention could turn to lesser but important alerts regarding things like financial and commercial interests.

Phase five - Coping and restoring: this would be about restoring normality and trying to recover damaged systems. The challenge would be to reach this phase as quickly as possible with the least sustained damage.

State of play

If even the security-heavy US is concerned about its grid, the same is likely to be true of most countries. I suspect many countries are not well-drilled to cope with sustained DDoS, especially given the fundamental weaknesses in DNS servers. Small countries are particularly at risk because they often depend on infrastructure that reaches a central point in a larger country nearby.

The UK, it should be said, is probably better placed than some countries to survive cyber warfare. It enjoys an independent grid and GCHQ and the National Crime Agency have helped to encourage some of the best private sector security operations centers in the world. Many countries could probably learn a great deal from it. Estonia, whose infrastructure was disabled for several days in 2007 following a cyber-attack, is now looking at moving copies of government data to the UK for protection.

Given the current level of international tension and the potential damage from a major cyber-attack, this is an area that all countries need to take very seriously. Better to do it now rather than waiting until one country pays the price. For better and worse, the world has never been so connected.

The Social Robots

Just imagine the brave new world where, there is nary a work to do, where you have all the time to stand and stare or devote to work that is more intellectual or spiritual - the "real" 'Brave New World'. Not too long ago, robots were giant, caged things, mainly found in automotive manufacturing lines. Social Robots was a new field of research pursued by the best and brightest in university research labs.

In the past few years, however, it seems that social robots have finally come of age. All of a sudden, the market is teeming with products. Some are distinctly humanoid.

The rise of social robots

Softbank Robotics' Nao, Pepper, and Romeo - all have a head and two arms. With their stylized designs, they deftly avoid the 'uncanny valley' of human-machine interfaces (realistic enough to look human, but non-human enough to look spooky).

Others are more subdued in their anthropomorphism. Blue Frog Robotics' Buddy sports an animated face on a screen, and scoots around on wheels. Jibo is yet more subtle in its ability to evoke humanity, with its stationary base and a head that can turn and nod.

What is a social robot, and what makes it special?

In her seminal paper, "Towards Sociable Robots, Professor Breazeall, inventor of Jibo, describes social robots thus: "(Social robots have) the ability to interact with people in an entertaining, engaging, or anthropomorphic manner."

The most noticeable quality in the interactions between a person and a social robot is the emotion. This can be partially achieved via a speech-based interface. For example, Amazon is working on an enhancement of Alexa, the virtual assistant that lives inside the Echo device, to help it understand emotion. China's Turing Robot goes a step further, and claims that its Turing Robot OS already understands emotion.

However, Dr. Breazeal's research has demonstrated that adding non-verbal cues via physical gestures can have a profound effect in increasing people's engagement, trust and confidence with the robot. In her Ted talk, she discussed how people interacted with a robotic diet-and-exercise coach significantly longer than they did with the exact same coaching program running on a computer. A disembodied voice coming out of a mobile device, or a cylinder on the kitchen counter, leaves a lot of engagement on the table.

A solution, or a gimmick?

So, social robots can be great interfaces. However, an interface, even one that embraces emotion, is only useful as long as it connects people to a solution that solves a problem. And therein lies the challenge. What do these robots do? What problems do they solve? And the reality is, no one has really made it past the cool factor, and arrived at a killer app that will take the mass market by storm.

Let's look at the Softbank Robotics product line. Pepper, Nao and Romeo are research platforms, designed for developers – not ordinary people. Pepper is the first consumer-facing solution that is designed to interact with end users out of the box. It has been deployed as a store attendant in a variety of retail settings. It is also being tested in Japanese homes as a companion robot.

But what is Pepper's job description in the store? A store representative can greet customers, help them find items, and ring up purchases. Pepper can greet customers. Is it addressing a real need, or is it a gimmick?

Household robots are not far away

As Pepper has some emotional intelligence, one could make the case that it can be a great companion robot in the home. Let's look at who needs companionship at home.

Children or elderly people come to mind. Children would gladly play with a robot. But a robot is no substitute for an adult. Any parent or caregiver of a young child knows this. And elderly people do need companionship. But what they crave is time with those they love, not companionship with a machine. Anybody who has cared for an elderly loved one knows this.

To their credit, Softbank Robotics' website makes it clear that the robot companion is meant to complement, not replace, people. However, it is not clear to what extent today's social robots can lessen the load for caregivers. The effectiveness of companion robots will improve in time, but there is work to do.

Last but not the least, there is the holy grail of service robots in the home. What about a robot helper that can clean and cook? Alas, the technology in sensing, navigation and manipulation is so far behind the real challenges of a messy household that a truly useful robotic helper is years away. The best that today's technology can offer is special purpose devices like the Roomba, or AI driven bar tending machines.

As a long time robot enthusiast, I can't wait to see robots become ubiquitous in the home. Right now, the interface is more advanced than what the robots can do. Only when the back end meets the front end in sophistication will we really start to see truly useful robots in the home.

Shared AI: Opportunity or Challenge

The labour impact of robots in the United States is crucial to the argument of whether it should be proliferated. Jason Furman, former chairman of the President's Council of Economic Advisers gave a speech on the "opportunities and challenges" presented by artificial intelligence. The potential challenges of artificial intelligence are, to put it in the terms of some recent economic policy conversations, problems posed by the rise of robots. The question at the heart of the debate is how concerned should we humans be about the impact of future technological change on our economy and society? Will we all be thrown out of work? Will new technologies make production so easy that economic growth absolutely booms? Or, perhaps more realistically, will something in between happen? The answer lies, to an extent, on how technological growth affects demand for labor in the United States and around the world.

First, a nod to Furman's point that a few more robots might be a good thing for the economy writ large at this point. Insofar as more artificial intelligence would boost productivity, the introduction of more "robots" would give the U.S. economy and those of other high-income countries a much needed boost in productivity growth. How much further "AI" innovation could deliver in an economy where the diffusion of innovation may be a major factor holding back productivity is up for debate. Yet most of the concerns about the rise of the robots are centered on distributional concerns.

The robots might boost productivity growth and therefore the growth of economic output, but will they displace large swaths of workers? Maybe permanently? This first fear—that robots will massively displace labor across manufacturing and services industries—is a concern because in the future additional capital investment in all manner of robots may well supplant labor rather than supplement it. If robots can essentially replace a large chunk of labor, then businesses will stop hiring workers and instead replace them with robots. Imagine an economy-wide version of robots on factory floors.

Past experience with technological change, Furman argues, shows that new technologies don't reduce demand for all labor, but rather shift the composition of demand for workers. Massachusetts Institute of Technology economist David Autor agrees, arguing that commentators overstate the ability of robots to substitute for labor and forget how capital acts as a complement to labor. The capital and labor complement each other, so that more capital accumulation or declining costs of investment raise the share of income accruing to labor.

Yet advances in artificial intelligence definitely change the kind of skills that labor needs to supplement the next generation of robots. This kind of deep technological change could reduce demand for, say, middle-skilled workers while boosting demand for less-skilled workers and more highly-skilled workers. The overall level of labor demand might not change, but the overall composition of the earnings distribution could well result in more income inequality. Of course, technology isn't the only thing that affects the distribution of income—labor market institutions such as union membership also play a significant role.

The extent to which future technological change affects the distribution of income will probably rest on how it impacts the overall demand for labor as well as the kind of skills that become more valuable in the future. There is some definitive evidence that overall demand for skilled labor is on the decline, but how long this recent trend continues and how much technology is responsible for it are open questions.

Impact of Robotics & Automation on Jobs

There is no need to worry about whether robots might start taking our jobs. It's already happening. Three of the world's 10 largest employers are already replacing tens of thousands of their workers with robots. And a recent report from Citi, produced in conjunction with the University of Oxford, highlights how increased automation could lead to greater inequality. 'If machines are capable of doing almost any work humans can do, what will humans do?'

The rise of robots and AI in the workplace seems almost inevitable now. Pretty much every startup now tends to include AI in some form or another. Moshe Vardi, a computer science professor at Rice University in Texas, told the American Association for the Advancement of Science : "We are approaching the time when machines will be able to outperform humans at almost any task. Society needs to confront this question before it is upon us: if machines are capable of doing almost any work humans can do, what will humans do?

A typical answer is that we will be free to pursue leisure activities. [But] I do not find the prospect of leisure-only life appealing. I believe that work is essential to human well-being."

However, let us look at scenario as it is unfolding:

Foxconn, a key manufacturing partner for Apple, Google, and Amazon, is the world's 10th largest employer and it has already replaced 60,000 workers with robots, according to a recent note written in part by analyst John Seagrimat CLSA.

Walmart, the third largest employer with 2.1 million workers is planning to replace its warehouse stock-checkers with flying drones that can scan miles of shelves in a fraction of the time.

And the US Department of Defense, the No.1 global employer, is already flying the world's largest fleet of unmanned aerial vehicles – drones, basically – in its various Middle East conflicts. The US Defense Department (DOD) has at least 7362 RQ-11 Ravens in operation.

More than five million individuals will lose their jobs by 2020 as part of the "fourth industrial revolution", which was also the theme of the 46th World Economic Forum (WEF). The jobs will be lost to robots because of a drastic change in the workplace that will be caused by artificial intelligence, according to a WEF study.

Robots will take over jobs that require more "narrow skills" such as administration or clerical work as employers start seeking new core skills such as critical thinking, emotional reasoning and "active listening". White collar workers will be affected the most, with about 4.8 million jobs in administration and clerical roles alone getting lost to robots. About 70% of the children studying in primary schools today will ultimately be working in jobs that do not exist as of today.

The five million number is the net figure. The total job losses will come to seven million which will be set off against two million jobs created, because of the revolution, according to the study, whose authors include Klaus Schwab, WEF founder, and Richard Samans, its former managing director. According to the authors, "Developments in genetics, artificial intelligence, robotics, nanotechnology, 3D printing and biotechnology, to name just a few, are all building on and amplifying one another. This will lay the foundation for a revolution more comprehensive and all-encompassing than anything we have ever seen."

Apart from creating job redundancies, the rise of the robots will affect gender equality at work. Women are expected to be more vulnerable to the job losses as a lesser number of them are at areas that will generate new jobs such as engineering, architecture, IT, software development and analytics. Over the next five years, while five women will lose their jobs, one woman will gain a job and when the same gets applied to men who are comparatively less vulnerable, they will see three jobs lost for every one gained.

That is likely the tip of the iceberg. 10 largest global employers show the potential for axing workers in favour of machines:

Health Service – with its massive army of doctors and nurses doing unrepetitive, unique tasks – looks like hostile territory for robots. The other nine are rich with rote, repetitive tasks that might be better performed by software.

77% of all Chinese jobs could go to robots

Foxconn's 60,000 robots are only a small fraction of its 1.3 million total workers. So we're still at the beginning of our robotic future. But few people are aware of the scale of the coming revolution in work.

Consider: The vision of mass unemployment created by robots isn't a guarantee, of course. The optimistic vision of a robotic future is one where our mechanical helpers allow us to work 4-day weeks by slashing the workload, leaving us to spend most of our time in the pub or in front of the

TV (let's be honest). New technology always tends to create more jobs than it initially destroys. But that happens in the longer run, as every robot worker will need a maker, a manager, and a maintenance person. The future is supposed to be a glorious place where robot butlers cater to our every need and the four-hour work day is a reality. But the true picture could be much bleaker.

Top computer scientists in the US warned over the weekend that the rise of artificial intelligence (AI) and robots in the workplace could cause mass unemployment and dislocated economies, rather than simply unlocking productivity gains and freeing us all up to watch TV and play sports.

Professor Vardi is far from the first scientist to warn about the potential negative effects of AI and robotics on humanity. Tesla founder Elon Musk co-founded a not-for-profit that shall advance "digital intelligence in the way that is most likely to benefit humanity as a whole" and Prof. Stephen Hawkins said: "The development of full artificial intelligence could spell the end of the human race."

The World Economic Forum also backed up Professor Vardi's fears with a report released in March 2017 warning that the rise of robots will lead to a net loss of over 5 million jobs in 15 major developed and emerging economies by 2020.

Could automation increase leisure time whilst also maintaining a good standard of living for everyone? The risk is that this increased leisure time may only become a reality for the under-employed or unemployed.
The report titled "Technology at Work: V2.0", concludes

that 35% of jobs in the UK are at risk of being replaced by automation, 47% of US jobs are at risk, and across the OECD as a whole an average of 57% of jobs are at risk. In China, the risk of automation is as high as 77%.

Most of the jobs at risk are low-skilled service jobs like call centers or in manufacturing industries. But increasingly skilled jobs are at risk of being replaced. The next big thing in financial technology at the moment is "roboadvice" – algorithms that can recommend savings and investment products to someone in the same way a financial advisor would. If roboadvisors take off it could lead to huge upheavals in that high-skilled profession.

Garlick writes: The big data revolution and improvements in machine learning algorithms means that more occupations can be replaced by technology, including tasks once thought quintessentially human such as navigating a car or deciphering handwriting.

Of course, these are theoretical risks – technology exists or is in within reach that means these jobs could be done by robots and machines, but it doesn't necessarily mean they will be. And the report is, in general, optimistic about the future of automation and robotics in the workplace. Citi says governments and populations are going to have to prepare for these changes, which are going to hit the world of work faster than technology advances have in the past.

The report predicts that many workers will have to retrain in their lifetime as jobs are replaced by machines. Citi recommends investment in education as the single biggest factor that could help mitigate the impact of increased automation and AI.

'Inequality between the 1% and the 99% may widen as workforce automation continues'. But within that recommendation the biggest issue associated with rising robotics and automation in the workplace is – inequality. Unlike innovation in the past, the benefits of technological change are not being widely shared – real median wages have fallen behind growth in productivity and inequality has increased.

The European Centre for the Development of Vocational Training (Cedefop) estimated that in the EU nearly half of the new job opportunities will require highly skilled workers. Today's technology sectors have not provided the same opportunities, particularly for less educated workers, as the industries that preceded them.

Not only is technology set to destroy low-skilled jobs, it will replace them with high-skilled jobs, meaning the biggest burden is on the hardest hit. The onus will be on low-earning, under-educated people to retrain for high-skilled technical jobs – a big ask both financially and politically.

The expanding scope of automation is likely to further exacerbate income disparities between US cities. Cities that exhibited both higher average levels of income in 2000 as well as the average income growth between 2000 and 2010, are less exposed to recent trends in automation. Thus, cities with higher incomes, and the ones experiencing more rapid income growth, have fewer jobs that are amenable to automation. Similarly, cities with a higher share of top 1 percent income earners are less susceptible to automation, implying that inequality between the 1 percent and the 99 percent may widen as workforce automation continues. In contrast, cities with a larger share of middle class workers also are more at risk of computerization.

Hence, new jobs have emerged in different locations from the ones where old jobs are likely to disappear, potentially exacerbating the ongoing divergence between US cities. Looking forward, this trend will require workers to relocate from contracting to expanding cities. And not only will the less well-off be forced to make the most changes in the robot revolution – reeducating and relocating – those that do retrain will be competing for fewer and fewer jobs.

Here's the Citi report: This downward trend in new job creation in new technology industries is particularly evident starting in the Computer Revolution of the 1980s. For example, a study by Jeffery Lin suggests that while about 8.2% of the US workforce shifted into new jobs during the 1980s which were associated with new technologies; during the 1990s this figured declined to 4.4%. Estimates by Thor Berger and Carl Benedikt Frey further suggest that less than 0.5% of the US workforce shifted into technology industries that emerged throughout the 2000s, including new industries such as online auctions, video and audio streaming, and web design.

The study suggests that new technologies are creating fewer and fewer jobs and it is likely that advances in automation and AI will destroy jobs at a much faster rate than it creates new roles.

According to Citi's report there will be 9.5 million new job openings and 98 million replacement jobs in the EU from 2013 to 2025. Roughly half of the jobs available in the EU would need highly skilled workers.

Yes, automation and robotics will bring advances and benefits to people – but only to a select few. Shareholders, top earners, and the well-educated will enjoy most of the benefits that come from increased corporate productivity and a demand for technical, highly-skilled roles.

Meanwhile, the majority of society – middle classes and, in particular, the poor – will experience significant upheaval and little upside. They will be forced to retrain and relocate as their old jobs are replaced by smart machines. But the pessimistic vision of the future is more inequality – cheap robots put the low-paid and middle classes out of work while reaping more and more profits for the factory masters and fat cats at the top of the chain. This phenomenon can also be seen in news publishing. The internet has created plenty of media jobs, but not as many as were destroyed in the print sector: '

The Citi report warned that advances in automation and robotics are likely to exacerbate the gulf between the 1% and the 99%. Jim Snabe, a board member at WEF, set out this binary vision at the DocuSign Momentum conference in London, telling the audience: There's no doubt that the opportunity for businesses to reinvent themselves. But the societal impact of that can be – will be – huge. There's a good and a bad version of that. The bad version is one where we bet everything on that digital infrastructure and yet it's not secure, we don't know what privacy is, and jobs get killed faster than we create new ones. Then this digital revolution will just be an ordinary revolution. People won't accept the fact that there's no longer a need for them.

The good version of this is one where we begin to steer the innovation towards a path where we use the opportunity to get independent limited resources – we stop climate change, we empower 4 billion people who are today not connected to the world.

The optimistic vision sounds very optimistic and, unfortunately for everyone, it looks like the pessimistic vision is so far making better progress than its rival – particularly in China. 'The bad version is one where ... jobs get killed faster than we create new ones'.

Two recent notes from CLSA and Bank of America Merrill Lynch draw attention to just how rapid the change is happening there. It a note on "Human Capital", BoAML says: "China has the fastest adoption of robots in the world and purchased over 57,000 industrial robots in 2014 (25% of global shipments). We believe China will witness a 25%+ CAGR in annual demand for new robots during 2014-18."

McDonald's, which employs 1.9 million people could even embrace the robot. McDonald's chief executive Ed Rensi told Fox Business that if minimum wage in the US rose, the fast-food chain would consider robots, according to CSLA's Seagrim. "It's cheaper to buy a $35,000 robotic arm than it is to hire an employee who is inefficient making $15 an hour bagging French fries," he told the broadcaster.

Rise of the new Luddites?

Seagrim starts his note by referencing the Luddites, the machine breakers of 19th century England, quoting the incredible fact that: "In 1811 more British soldiers were deployed fighting Ned Ludd and his army of unemployed textile workers in the North of England than were engaged fighting Napoleon in Spain!" Clearly machines stealing jobs is not a new phenomenon.

Despite the pace of change, Seagrim believes the outcome won't be as violent as the Luddite era: "In a human-machine study conducted by an MIT professor at a BMW factory, it was shown that teams made of humans and robots collaborating efficiently can be more productive than teams made of either humans or robots alone. The cooperative process reduced human idle time by 85%." That may be great in theory but as WEF's Snabe points out, humans need a concerted and conscious effort from businesses to include them rather than just focusing on the bottom line.

The societal impact is huge and I think we have an opportunity and an obligation as business leaders to think in that direction rather than simply optimizing for the next quarter. What are we going to do to make sure there are relevant jobs – they will be different jobs – for the future. We talk about the fourth industrial revolution – the second one was electricity. Well, there are still 1.4 billion people without access to electricity. Imagine we use our innovation capacity to solve this, we bring 4 billion people online – this is the future that we need.

We find in Canada a clear evidence of loss of jobs. More than 40 per cent of the Canadian workforce is at high risk of being replaced by technology and computers in the next two decades, according to a new report. The Brookfield Institute for Innovation + Entrepreneurship at Toronto's Ryerson University said in its report that automation previously has been restricted to routine, manual tasks. However, breakthroughs in artificial intelligence and advanced robotics now mean that automation is moving into "cognitive, non-routine tasks and occupations, such as driving and conducting job interviews."

The report said the top five occupations — in terms of number of people employed in them — facing a high risk of automation are:

- **Retail salesperson.**
- **Administrative assistant.**
- **Food counter attendant.**
- **Cashier.**
- **Transport truck driver.**

The institute put a 70 per cent or higher probability that "high risk" jobs will be affected by automation over the next 10 to 20 years, and it said workers in the most susceptible jobs typically earn less and have lower education levels than the rest of the Canadian labour force.

Jobs deemed to be at a low risk of being affected by automation — having a less than 30 per cent chance — are linked to high skill levels and higher earnings, such as management and jobs in science, technology, engineering and math (STEM).

The institute also said workers in the jobs deemed at high risk in the study are disproportionately between 15 and 24 years old , while workers in lower risk jobs tend to be "prime-aged workers," between 25 and 54. "Canada's younger and, to a lesser extent, older populations are more likely to be vulnerable to the effects of automation," the study said.

I visualize; the same situation in South Asia. More than half of workers in five Southeast Asian countries are at high risk of losing their jobs to automation in the next two decades; an International Labour Organization study found, with those in the garments industry particularly vulnerable. About 137 million workers or 56 percent of the salaried workforce from Cambodia, Indonesia, the Philippines, Thailand and Vietnam, fall under the high-risk category, the study showed. "Countries that compete on low-wage labor need to reposition themselves. Price advantage is no longer enough," said Deborah France-Massin, director for the ILO's bureau for employers' activities. The report said workers have to be trained to work effectively alongside digitalized machines.

Southeast Asia is home to more than 630 million people and is a hub for several manufacturing sectors, including textiles, vehicles and hard disk drives. Of the 9 million people working in the region's textiles, clothing and footwear industry, 64 percent of Indonesian workers are at high risk of losing their jobs to automation; 86 percent in Vietnam, and 88 percent in Cambodia face the same risk.

Garment manufacturers in Cambodia, who take orders from retailers such as Adidas, Marks and Spencer and Wal-Mart Stores Inc, employ about 600,000 people. Neighboring Vietnam is seeing record investment in its footwear and textiles industries, due to new free-trade pacts with major markets. It is the second-largest garment supplier behind China to the United States.

The United Nations agency said technologies including 3D printing, wearable technology, nanotechnology and robotic automation could disrupt the sector. "Robots are becoming better at assembly, cheaper and increasingly able to collaborate with people," the ILO said.

The textiles, clothing and footwear sector is at the highest risk of automation out of five industries analyzed in the study, including automotive and auto parts, electrical and electronics, business process outsourcing and retail.

In the automotive and auto parts industry, more than 60 percent of salaried workers in Indonesia, and over 70 percent of those in Thailand face the risk of their jobs being displaced. Southeast Asia's automotive sector, the seventh-largest producer of vehicles in 2015 globally, employs more than 800,000 workers, the report said.

Known as the "Detroit of Southeast Asia", Thailand is a regional production and export hub for the world's top carmakers. The auto sector accounts for around 10 percent of Thai GDP and employs a 10th of its workers in manufacturing.

In March, a robot Spenser started escorting lost travelers to their gates at Amsterdam's Schiphol Airport. That same month, robot hosts welcomed visitors to big trade shows, while the Hilton McLean Virginia in Washington, D.C, welcomed Connie the robotic concierge. Then there's the robotic receptionist at Belgium's Ghent Marriott. Not only can Mario speak 19 languages, he does so while dancing to Michael Jackson.

While these may have a certain novelty factor for now, it's behind the scenes where artificial intelligence will really shake up the hospitality industry. According to the EU, the market for service robots will reach €100 billion a year by 2020.

There are two elements to robotics. There's the side you can see and touch, and then there's the processing that happens behind the scenes. Similarly, the goals of the technology are twofold. On the one hand to improve operations, and on the other to control costs, through increased efficiency, for example, or higher sales. US-based fast-food restaurant franchise Zaxby's have been using robotics for this purpose. Known as Hyperactive Bob, the unseen robot reduces waste and waiting times by scanning the parking lot for incoming cars while taking into account various data such as time of day and amount of hot food ready at a given moment.

"The algorithmic system senses demand based on customer arrivals and makes predictions that feed into planning," explains Pemberton. "This allows staff to adjust the timing of the cooking process, and respond to specific directions that speed up service, and thus turnover."

A fledgling robotics marketplace

Currently, front-of-house robots are able to carry out rudimentary functions, such as checking guests in or recommending local restaurants. "You wouldn't replace your concierge with a Connie—not just yet anyway," says Pemberton. "Hotels should be trying to engage people with a product that works alongside the human team, as opposed to replacing it."

Nua Robotics, for example, have produced a suitcase that follows you around. "It works using Bluetooth and cameras to detect where you are," explains Pemberton. "Hotels could adopt the technology to track guests' luggage." Similarly, Savioke's Relay can learn the layout of the room and deliver items to guests on demand. Though not quite ready to offer full room service (the Relay can't carry hot food), the autonomous android is already being employed by hotels, like Starwood's Aloft Group, who have named their new staff member Botlr.

Such robots, whether they're making deliveries or welcoming guests, have been well received by the public. But it's the crossover of front-facing robots and back-end artificial intelligence that has the potential to transcend the novelty factor. "AI is where we're going. Once we've perfected that side of things, it'll be a huge boon to this industry."

Robots meet Big Data

Hilton's Connie is among the most advanced products from this perspective. Named after the hotel chain's founder Conrad, Connie is the hospitality industry's first concierge powered by IBM Watson, which is described as "a technology platform that uses natural language processing and machine learning to reveal insights from large amounts of unstructured data."

Watson's powerful AI capabilities means Connie can scale human staff's expertise by surfacing answers to specific requests. The robot is also an expert at natural language processing, able to understand idioms and draw inferences from human speech. What's more, it learns, adapts and gets smarter.

"With all IBM's customer relationship management solutions, it's not going to be too long before they add the ability for robots to recognize people as they come into the hotel," says Pemberton. "It could connect to the guest's smartphone via Bluetooth or use facial recognition."

An intelligent interface like this can also collect and structure data about guest behavior and preferences, and begin to build up a profile that helps hotels deliver a truly personalized and seamless customer experience. "Front-facing robots play an important role in data gathering, and can then make connections. For instance, they can store a query and intelligently offer relevant information to a guest in future," says Pemberton. Crucially, such data can be amassed in a centralized location and delivered on demand locally; so the preferences of a guest staying in a hotel in Miami will be captured and reapplied when she checks into a hotel belonging to the same chain in say, London.

Ritz-Carlton's Mystique system employs a similar approach, though the data is predominantly gathered by humans at the moment. "Whenever we discover or are alerted to a guest's preferences, we put them into Mystique, and Mystique talks to all the properties within Ritz-Carlton," Diana Oreck, vice president of the hotel chain's Leadership Center told Forbes. "As a guest, you can go from Ritz-Carlton to Ritz-Carlton around the world, and we will know and be able to deliver what you like."

Human and robots working together

While the potential for robotics to transform the industry cannot be underestimated, human staff can take comfort in the fact that they won't be replaced anytime soon. Indeed, "robots like Connie can augment cognitive capacities and help staff do their job," says Pemberton. "Your friendly human concierge may not be able to moonwalk as well as Mario, but he possesses the skills and nuance that a robot just can't replace."

However, this is just the beginning – and the concept of artificial intelligence is evolving rapidly. "There are so many untapped possibilities, with potential benefits for consumers and the hotel industry alike," Pemberton says. "The future looks very interesting indeed."

The Economic and Social Impacts of Robots

Robots are just one of the latest stages of technological progress. The number of robots being used by businesses to boost productivity has increased rapidly in recent years. And there is no reason to believe that this pace of robotization will begin to slow any time soon.

On the contrary, as the cost of robots continues to fall while their capabilities go up, and with the robot density in most industries still relatively low, the International Federation of Robotics (IFR) anticipates that yearly robot installations will continue to grow at double-digit rates for the time being.

Rising inequality and slow productivity gains may be the main economic challenges of the 21th century. And the increased use of robots should affect both of these developments – positively as well as negatively. While empirical literature about the impact of robots is still in its infancy, there is now a growing number of studies which begin to support the notion that they lift productivity, wages and even total labor demand, but mostly benefit higher-skilled workers. With the increased use of robots, computers and other machines, the latest round of technological progress now largely comes at the expense of middle and low-skilled, and wage workers.

According to these studies, the productivity impact of robots is already comparable to the contribution of steam engines. And while still lagging behind the impact of ICT (information and communications technology), one has to keep in mind that the total value of ICT capital by far exceeded that of current robot services. Some of the productivity gains from robot densification are shared with workers through higher wages.

The issue is, however, that different income and skill groups do not benefit to the same extent, which means that robotization further adds to income inequality. To allow for a broader share of the population to reap the benefits of this technological progress, two sets of actions should be taken.

Skills and Education

We need to rethink our education system. As robots and machines are capable of taking over a growing number of tasks, humans have to focus on their comparative advantages, including non-cognitive skills. In addition, advanced countries (notably the US) have to halt and reverse the trend that the quality of student education is primarily determined by parents' income and wealth, as this unequivocally amplifies the negative inequality spiral. But even if politicians do adopt the necessary changes to the education system, increased technological progress will most likely still lead to growing income inequality, as people have different skills, as well as different financial conditions.

Spreading the Ownership

Due to this inequality, there is a growing need to reallocate income from rich to poor and/or from owners to workers. In theory, there are three possibilities to try to partly offset or mitigate the ongoing decline in labor's share of income:

1) Higher wages through collective bargaining or minimum wages.

2) Redistribution of wealth and income through tax-and-spend policies.

3) Spreading the ownership of capital to ensure a more equitable distribution of robotic rents.

The first two options have been the traditional ways to redistribute profitability and income gains, and they will certainly be used again this time. There are, however, tight limits to what can be achieved through them. Indeed, if robots compete with low and medium-skilled workers, raising (minimum) wages would only accelerate and intensify the substitution of labor with capital.

One of the most promising solutions to the long-term challenge posed by machines substituting for labor. Unless workers earn income from capital as well as from labor, the trend toward a more unequal income distribution is likely to continue, and the world will increasingly turn into a new form of economic feudalism. We must widen the ownership of business capital if we hope to prevent such a polarization of our economies. To our mind, employee-ownership is one of the most promising solutions to the long-term challenge posed by machines substituting for labor. It allows workers to earn income from labor as well as from capital.

Artificial Intelligence in 2030

Artificial intelligence (AI) has already transformed our lives: from the autonomous cars on the roads to the robotic vacuums and smart thermostats in our homes. Over the next 15 years, AI technologies will continue to make inroads in nearly every area of our lives, from education to entertainment, health care to security.

The question is, are we ready? Do we have the answers to the legal and ethical quandaries that will certainly arise from the increasing integration of AI into our daily lives? Are we even asking the right questions?

Now, a panel of academics and industry thinkers has looked ahead to 2030 to forecast how advances in AI might affect life in a typical North American city and spark discussion about how to ensure the safe, fair, and beneficial development of these rapidly developing technologies.

"Artificial Intelligence and Life in 2030" is the first product of an ongoing project hosted by Stanford University to inform debate and provide guidance on the ethical development of smart software, sensors, and machines. Every five years for the next 100 years, the AI-100 project will release a report that evaluates the status of AI technologies and their potential impact on the world.

"Now is the time to consider the design, ethical, and policy challenges that AI technologies raise," said Grosz. "If we tackle these issues now and take them seriously, we will have systems that are better designed in the future and more appropriate policies to guide their use."

"We believe specialized AI applications will become both increasingly common and more useful by 2030, improving our economy and quality of life," said Peter Stone, a computer scientist at the University of Texas, Austin, and chair of the report. "But this technology will also create profound challenges, affecting jobs and incomes and other issues that we should begin addressing now to ensure that the benefits of AI are broadly shared."

The report investigates eight areas of human activity in which AI technologies are already affecting urban life and will be even more pervasive by 2030: transportation, home/service robots, health care, education, entertainment, low-resource communities, public safety and security, employment, and the workplace.

Some of the biggest challenges in the next 15 years will be creating safe and reliable hardware for autonomous cars and health care robots; gaining public trust for AI systems, especially in low-resource communities; and overcoming fears that the technology will marginalize humans in the workplace.

Issues of liability and accountability also arise with questions such as: Who is responsible when a self-driven car crashes or an intelligent medical device fails? How can we prevent AI applications from being used for racial discrimination or financial cheating?

The report doesn't offer solutions but rather is intended to start a conversation between scientists, ethicists, policymakers, industry leaders, and the general public.

Grosz said she hopes the AI-100 report "initiates a century-long conversation about ways AI-enhanced technologies might be shaped to improve life and societies."

Robots not a Replacement for Humans

Automation has become a concern not just for blue-collar manufacturing workers, but also for white-collar workers and even for professionals. New computer programs, some using artificial intelligence, are taking over the tasks of bookkeepers, bank tellers, clerks, and others. Some see this replacement causing technological unemployment and a slow recovery from the Great Recession. 47% of total US employment is...potentially automatable over...perhaps a decade or two. The view that computer automation has been causing and will increasingly generate major unemployment has prompted calls for new policies such as a minimum basic income.

But has computer automation actually been generating a large net loss of jobs? Unfortunately, much of the popular discussion of automation has not benefited from either rigorous economic analysis or empirical evidence.

Some basic economics of automation

It is important to begin with a clear understanding of what automation is and how it affects jobs. With automation, machines perform part or all of an occupational task, reducing or eliminating the human labour needed to perform that task. But this is not the only way that new technology can disrupt the workforce. New technology can make products obsolete. For example, the automobile eliminated jobs for carriage makers, although it also created jobs for auto-body makers. Technology can also change work organisation. For example, communication technologies facilitate decentralization, outsourcing, and offshoring, shifting work from one group of workers to another. Self-service technologies (e.g. the airline ticket kiosk) shift work to consumers. Information technology can facilitate new markets (e.g. Airbnb, Uber). Although some technological change can be disruptive and eliminate jobs for some workers, there is no particular reason to expect them to create large job losses overall; new jobs are created while old ones are eliminated. Automation, on the other hand, might cause net job losses because machines reduce the human labour needed to produce a unit of output.

Also, much of the discussion concerns human jobs being completely taken over by machines. But in fact, most automation is partial—only some tasks are automated. For example, despite extensive automation since 1950, it appears that only one of the 270 detailed occupations listed in the 1950 Census was eliminated thanks to automation – the job of elevator operators. Many others, however, were partially automated.

This distinction is important because it implies very different economic outcomes. If a job is completely automated, then automation necessarily reduces employment. But if a job is only partially automated, employment might actually increase. This is true even if the job is mostly automated. The reason has to do with basic economics. For example, during the 19th century, 98% of the labour required to weave a yard of cloth was automated, yet the number of weaving jobs actually increased. Automation drove the price of cloth down, increasing the highly elastic demand, resulting in net job growth despite the labour saving technology.

Similar demand responses are seen with computer automation. Consider, for example, the effect of the automated teller machine (ATM) on bank tellers. The number of fulltime-equivalent bank tellers has grown since ATMs were widely deployed during the late 1990s and early 2000. Why didn't employment fall? Because the ATM allowed banks to operate branch offices at lower cost; this prompted them to open many more branches (their demand was elastic), offsetting the erstwhile loss in teller jobs.

Of course, partial automation can also decrease employment in an occupation. If demand is inelastic, then growth in demand will not offset job losses. Also, automation can lead to substitution of one occupation for another within firms and industries. For example, there are fewer telephone operators now, but more receptionists; there are fewer typesetters, but more graphic designers, and desktop publishers. Graphic designers using computers became more productive than typesetters; so, automation facilitated the shift of work from typesetters to graphic designers.

Estimates of employment demand growth

Taking these considerations into account, I assume a simple model of occupational demand across industries that allows for changing demand and inter-occupation substitution within industries. As the key independent variable, one can measure the extent of computer use by workers in each occupation and industry. The occupations that use more computers will have a higher degree of task automation, all other factors being equal. The dependent variable is the relative growth of employment in occupation-industry cells.

The estimates contradict popular assumptions about the impact of computer automation. First, computer-using occupations tend to grow faster, not slower. At the sample mean, computer use is associated with a 1.7% increase in occupational employment per year. In other words, the bank teller example may be typical rather than exceptional.

Second, there is a strong substitution effect between occupations. Occupations tend to have declining growth to the extent that other occupations in the same industry use computers. That is, the story is not about machines replacing humans; rather it is one of humans using machines to replace other humans, as graphic designers with computers replaced typesetters.

The substitution effect largely offsets the growth effect. Counting both, at the sample mean, computer use is associated with positive employment growth but the effect is small, 0.45% per year. This association is not necessarily causal—perhaps some other factor caused computer-using occupations to grow. But this finding does show that computer automation is not associated with major job losses.

Computer automation and inequality

Nevertheless, computer automation is associated with major workforce dislocation. While automation does not appear to have a major effect on overall employment, automation is associated with substantial job losses for some groups of occupations and job gains for other occupations. In particular, low-wage occupations tend to lose jobs while high-wage occupations gain . High-wage occupations use computers more intensively, allowing them to substitute for work done by low-wage occupations.

This disparity could contribute substantially to economic inequality if workers in low wage occupations cannot easily transfer to high wage occupations. Low-wage workers, for instance, might not get opportunities to work with computers or might not have the necessary skills. Occupations that use computers more heavily have had growing dispersion of within-occupation wages—workers who acquire new skills earn more, but not all workers have the opportunity or ability to learn. Also, computer-using occupations tend to employ increasing shares of college educated workers, even in occupations such as bank teller that do not require college degrees.

Computer automating tasks do not imply that occupations that use computers will necessarily suffer job losses. In fact, computer-using occupations have had greater job growth to date. Instead, it is the occupations that use few computers that appear to suffer computer-related job losses.

The notion that computer automation necessarily leads to major job losses ignores the dynamic economic response to automation, a response that involves both changing demand and inter-occupation substitution. Of course, the recent experience does not necessarily predict the future and new artificial intelligence technologies might have a different effect. Indeed, even though past technologies, such as automated weaving, initially created many jobs, demand elasticity eventually declined, and then further technological gains led to job losses. Computer automation may create job losses in the future.

But focusing on that future problem is a poor guide for today's policy. The evidence suggests that while computers are not causing net job losses now, low wage occupations are losing jobs, likely contributing to economic inequality. These workers need new skills in order to transition to new, well-paying jobs. Developing a workforce with the skills to use new technologies is the real challenge posed by computer automation.

Human Shall Thrive in the Machine Age

The world is not just rapidly changing, it is being dramatically reshaped. It is being reshaped faster than individual humans and the institutions are *yet* able to respond. Recent technological advances and disruptions have generated a world that operates so differently that we struggle to comprehend its meaning and adapt to the circumstances it presents to us. This new world poses profound challenges for organizations of all kinds as they try to cultivate resilience and simultaneously determine a source of growth.

In January 2016, World Economic Forum Executive Chairman Klaus Schwab recognized this challenge, referring to this economic and cultural upheaval as the Fourth Industrial Revolution. He particularly called out the fact that the boundaries dividing the physical and digital worlds are growing less defined, and the difference in abilities that so clearly separated man from machine is quickly eroding.

Take, for instance, the recent surge in popularity of the augmented reality mobile game, *Pokeman Go* – which attracted over 25 million users in the span of a few weeks. Its success offers just a single recent example, among many, of how our natural and virtual worlds are melding into one another. And as machines become more thoughtful and intellectually nimble, humans must not only contend with the prospect of future joblessness but also the deeper implications of what it means to be replaceable as a species. The rapid advances in artificial intelligence force us to grapple with a question that is both philosophical and deeply personal: "What makes me special?"

Meaning of being human in the age of machines

We are in the midst of a revolution, one that dwarfs the so-called industrial revolutions that preceded it. What we are experiencing today bears striking similarities in size and scope to the Scientific Revolution of the 16th century. The discoveries of Copernicus and Galileo, which challenged our understanding of the world around and beyond us, inspired others to ask deep questions about the nature of humanity and how societies should be organized and governed. The Scientific Revolution disrupted the way the human race thought about itself. We now have a chance to embrace today's revolution for what it is: a powerful, defining moment to rethink what it means to be human. Our present revolution is not only one where we will need to rethink the nature and structure of our industries and institutions, but also one where we will need to create new systems that put humanity at the centre and come to grips with the transformative implications of such a change.

We must summon the courage to answer what it means to be a person in the age of intelligent machines, and what makes organizations that are comprised of human beings — governments, corporations, or NGOs — human. Professor Schwab put this challenge into stark relief last October when he asked: Will the Fourth Industrial Revolution have a human heart"? This presents both a daunting challenge and an extraordinary opportunity. As Professor Schwab wrote: "the Fourth Industrial Revolution may indeed have the potential to 'robotize' humanity and thus to deprive us of our heart and soul. But as a complement to the best of human nature — creativity, empathy, stewardship — it can also lift humanity into a new collective and moral consciousness based on a shared sense of destiny."

In a world that increasingly defers to the cognitive dexterity of artificial intelligence, only the human heart can animate morality, imagination, ethics, compassion and empathy. The human heart exhibits consciousness of others and designs technology whose aim is human advancement and elevation, not destruction.

A human economy

Over the course of the 20th century, the mature economies of the world evolved from being *industrial* economies to *knowledge* economies. Now we are at another watershed moment, transitioning to *human* economies. We must now work to put our humanity at the center of all of our endeavors. We can do this by cultivating our most unique human quality: our ability to pause. When you hit the pause button on a machine, the action stops. But when humans pause, that's when true, reflective work begins.

Empathy, compassion, imagination, and other forms of elevated human behaviour cannot exist without pausing. By taking a step back from what is going on in our day-to-day lives, we free ourselves to reconnect to our deepest values and concerns, and face problems with integrity, courage, and humility. It is in the pause that we can reflect deeply about our challenges. By pausing, we reconnect with our deepest humanity and our source of meaning. We rethink our assumptions and models and reimagine new institutions and pass it forward. This approach allows us to view our problems in entirely new ways.

As traditionally high-skill jobs are increasingly outsourced to machines, humans must develop their deeper human capacities and qualities. In a twist of irony, the overwhelming advances in technology of the past few decades have made clear the critical need for humanistic education. As Microsoft CEO Satya Nadella recently wrote, "To stay relevant, our kids and their kids will need empathy … perceiving others' thoughts and feelings…. Ethics and design go hand in hand."

Humans and machines can complement one another, as long as we take care to consider things like empathy and ethics when developing artificial intelligence. As Nadella wrote, "if we've incorporated the right values and design principles, and if we've prepared ourselves for the skills we as humans will need, humans and society can flourish."

Machines will be able to carry out their assigned tasks, but only humans will be able to redefine, transcend and elevate those tasks. Machines can be measured according to whether they've met expectations. But humans are uniquely capable of focusing on how we collaborate, how we build trust, and how we innovate, recalibrating their behaviour as changes arise.

Thriving in the Fourth Industrial Revolution

So, how can business leaders and organizations create the conditions that will allow people to embrace their most uniquely human abilities? *The HOW Report*, a new cross-industry statistical analysis of 16,000 employees at companies spanning 17 countries, offers one way forward. It shows that self-governing organizations, those that are purpose-driven and give people the autonomy and flexibility to do their best work, are poised to achieve the best results in this new working world.

The report classifies organizations into three models: "blind obedience," "informed acquiescence", and "self-governing".

Blind obedience organizations operate via command-and-control principles and policing. Employees are expected to follow the rules and are punished if they fail to act within specified parameters. Such organizations place little emphasis on building relationships among colleagues, with customers, or within society. In a sense, blind obedience firms treat people like machines.

Informed acquiescence organizations are rules-based and process-driven. They operate through hierarchy and policy. They seek to motivate employees through performance-based rewards and punishments. While these organizations often have long-term goals, they also set them aside in the face of short-term challenges. These organizations often treat people like complex, adaptive machines.

Self-governing organizations, meanwhile, put humanity at the centre. Such organizations are propelled by the pursuit of significance and are led via moral authority, not hierarchical authority. Self-governing firms focus on long-term goals and achieving sustainable performance. They are inspired by shared values and ethics, not policies and dictates. Unlike blind obedience and informed acquiescence, two hallmark organizational styles of the Industrial Revolution, self-governance is a uniquely human capacity. Machines are capable of functioning in blind obedience and informed acquiescence organizations. But only humans can operate in a self-governing environment. Simply put, self-governing companies are the most human of all kinds of organizations.

Why self-governing companies outperform all others

The report shows that self-governing companies outperform those that follow the other two organizational models on several accounts. They are three times as likely to achieve high levels of performance as their blind obedience rivals, as measured by growth in market share, customer satisfaction, innovation, business sustainability, and employee engagement.

When Bill Gates developed the *Windows Operating System*, he created an environment in which several of Microsoft's applications could work together. Today, we must create a *Human Operating System*, the kind that will give elevated behaviours – courage, compassion, and creativity – room to grow and flourish. And self-governing organizations are best positioned to achieve this.

Building Robots with Human Emotions

I recently read an article called "Robots with Heart". In the piece, I found out a startlingly new idea of work incorporating an 'empathy module' into robots in order for them to better serve the emotional and physical needs of humans. While many have offered ideas on how we might apply these empathetic robots to medical or other applications, some objected to the very idea of making robots recognize and empathize with human emotions. One scientist opined that, as emotions are what make humans human, we really should not build robots with that very human trait and let them take over the care-giving jobs that humans do so well. On the other hand, there are others who are so enthusiastic about this very idea that they ask me, "If robots are intelligent and can feel, will they one day have a conscience?"

Perhaps it is important for us to understand what it means by robot intelligence and feeling. It is important for us to understand, first of all, how and why humans feel.

Human emotion

What is the role of emotion in the evolution of our species? Research has shown that humans bond with other humans by establishing a rapport. The survival of a species depends on that bonding, and much of this bonding is enabled by emotion. We also signal our intent with emotion.

Our feelings and emotions are triggered by stimuli, either external or internal (such as a memory), and manifest themselves in terms of physical signs – pulse rate, perspiration, facial expressions, gesture, and tone of voice. We might cry or laugh, shudder in disgust, or shrink in defeat. Unlike our language, much of these emotions are expressed spontaneously and automatically, without any conscious control. We learn to recognize emotions in other human beings from birth. The gentle humming of a lullaby soothes babies even before they are born. They respond to the smiling face of a parent at birth and are certainly capable of expressing their own emotions from day one.

Machine emotion

Industry robots build our cars and our smartphones. Rehabilitation robots help people walk again. Machine teaching assistants can answer student questions. Software programs can write legal documents. They can even grade your essays. Software systems can write stories for newspapers. An Artificial Intelligence (AI) program just beat a human at *Go*, known to be the most complicated board game. IBM Watson beat human champions at *Jeopardy*. Machines can even paint to the point of fooling humans into believing the result to be that of a professional human artist. Machines can compose music. Robots can obviously be built to be stronger, faster, and smarter than humans in specific areas. But do they need to feel like we do?

In early 2016, IBM team announced the first known system that can recognize a dozen human emotions from tone of speech instantaneously and in real-time. Prior to this work, recognizing emotions from tone of voice would incur some delay in processing time due to a procedure called 'feature engineering', a delay that is unnatural in a human-robot communication scenario. To understand how we achieved this, we need to understand machine learning.

Every robot is run on a hardware platform driven by software algorithms. An algorithm is designed by humans to tell the machine how to respond to certain stimuli for example, or how to answer a question, or how to navigate around a room. Much like an architect building a house, an AI engineer looks at the whole picture of what task the machine is supposed to achieve, and builds software 'blocks' to make it achieve that task. What is called programming is simply the implementation of the codes that realizes these blocks. One of the most important blocks is machine learning – algorithms that enable machines to learn and simulate human-like responses, such as a chess move, or to answer a question.

What has fueled real breakthroughs in artificial intelligence has been machine learning. Instead of being programmed to respond in certain, predictable ways, machines are programmed to learn from large amounts of real-world examples of stimuli-responses. If a machine looks at tons of cat pictures labelled 'cat', it can use any one of the many machine learning algorithms to recognize a cat from any unseen picture. If a machine looks at trillions of websites and their translations, it can learn to approximate translation in the manner of *Google Translate*.

A critical part of machine learning is to learn the representation of the characteristics, called features, of the physical input. A cat is represented by its contour, edge, facial and body features. The frequency components of the audio represent speech input. Emotions in speech are represented by not just the pitch, but the chroma, the tempo, the speed, of that voice. Machine learning needs to first perform feature engineering to extract these characteristics. For tone of voice, feature engineering typically extracts 1000-2500 characteristics from the input audio, and this process slows down the whole emotion recognition process. These thousands of features are carefully designed by humans and each of them requires processing time.

Recent breakthroughs in neural networks, aka deep learning, enabled by both machine speedup and massive amounts of data for learning, have led to vast improvements in machine learning. To start with, some deep learning methods, such as convolutional neutral networks (CNN), can automatically learn the characteristics during the learning process, without an explicit and delayed feature engineering process or human design. This is perhaps the most important contribution of deep learning to the field of AI.

Coming back to our emotion recognition system from tone of voice, what we did is replace feature engineering and classifier learning by a simple convolutional neural net, which learns just as well, if not better than,as classical machine learning approaches, and is much faster, because it does not require an explicit and slow feature engineering process. Similarly, facial expression recognition can be done in real-time with a CNN.

In addition, researchers are working to enable robots to express emotions - changing the pitch of its machine voice, using dozens to hundreds of tiny motors to control the synthetic facial muscles. The androids Sofia and Erica are two examples of humanoid robots with facial expressions.

Human-Robot Bonding

We are already shifting into the mode of the Fourth Industrial Revolution. And it seems that technology is poised to replace humans in many areas. Skills that took years, maybe decades, to acquire seem to become obsolete overnight. Most of the population are not aware of the pace of progress made in AI and robotics prior to the current torrent of publicity, and are extrapolating what they see today to predict that robots will take over in 30 years, or 50 years. There has been a lot of anxiety in the society regarding the question of if and when will robots 'take over' from humans.

Truth is, this kind of prediction has been around for a long time. It happened during all previous industrial revolutions, when people feared steam engines or that computers would render humans redundant. What has always happened is that people simply learned different skills to manage these machines, and more.

Nevertheless, with more applications of AI and robots, a new kind of relationship between humans and machines needs to evolve. For humans to be less fearful of and to trust a walking, talking, gesturing and weight-carrying robot, we need to have mutual empathy with the robot. What sets a robot apart from mere electronic appliances is their advanced machine intelligence – and emotions. To understand the cry of a baby, or the painful groan in a patient's voice, is critical to home care robots. For robots to be truly intelligent, they need to 'have a heart'.

Will robots be conscious?

If a robot develops analytical skills, learning ability, communication, and even emotional intelligence, will it have a conscience? Will it be sentient? Can it dream?

The above-mentioned neural networks, unlike other machine learning algorithms, remind people more of our own brains. Neural networks can even generate random, dream-like images, leading some to believe that even robots can dream.

The real question is do we understand what makes us humans sentient? Is it just the combination of our sensory perception and the thinking process? Or is there more to it? AI researchers cannot answer this question, but we do believe that to make 'good' robots we have to teach them values – a set of decision-making rules that follow our ethical and moral norms. With the expansion in robot intelligence, teaching values to machines will become as important as teaching them to human children. Our next challenge would be to enable automatic-machine learning of such values – once they have the prerequisite emotional recognition and communication skills.

How Human should be Robot Companions?

What would your ideal robot be like? One that can change nappies and tell bedtime stories to your child? Perhaps you'd prefer a butler that can polish silver and mix the perfect cocktail? Or maybe you'd prefer a companion that just happened to be a robot? Certainly, some see robots as a hypothetical future replacement for human care givers. But a question roboticists are asking is: how human should these future robot companions be?

A companion robot is one that is capable of providing useful assistance in a socially acceptable manner. This means that a robot companion's first goal is to assist humans. Robot companions are mainly developed to help people with special needs such as older people, autistic children or the disabled. They usually aim to help in a specific environment: a house, a care home, or a hospital.

At the beginning of the 20th century, one of the first pieces of technology designed to help in a household environment was the vacuum cleaner. Since then, technology has transformed the home. Nowadays, we even have a robot that can cook. The chef robot was developed by Moley Robotics, a start-up company that won the 2015 Asia Consumer Electronics Show. The robot is said to be able to cook 2,000 different meals.

Pepper, the latest robot from Aldebaran Robotics, is a good example of a humanoid robot companion. It can provide assistance in making choices, detect human facial expressions and communicate with people. Pepper can adapt its behaviour depending on its perception of a person's mood, and in this sense we can say that Pepper cares for people. At the moment, only research institutes and Japanese residents can acquire a Pepper robot. The robot costs around £8,710 for a Japanese customer.

Companion robots can take the form of pets, too. Paro is a robotic seal developed to provide comfort to old people. And rather than taking care of you, this robot has to be taken care of. It is how Paro provides emotional support.

Sometimes people get attached to robots that are not actually made for companionship. Take Roomba, for example, the intelligent vacuum cleaner. People wanted to become tidier in order to allow the vacuum cleaner to run smoothly.

Although many of these robots show some form of initiative and encourage people to interact with them, many are responsive rather than active; in other words, the robot waits for a human request before acting.

Should robots be more 'human'?

Thanks to progress in Artificial Intelligence and technology, we can now develop more intelligent systems that are capable of acting very much like a human. Last year, a few of them were presented to the public: such as Nadine, the robot receptionist; Yangyang, the singing robot, and Aiko Chihira, the robot that can communicate in sign language.

Although the popular and controversial Turing Test is used in AI to measure whether a machine is as intelligent as a human, it is a very different thing when it comes to robots, since robots are also expected to act intelligently. There is not yet a standardized test to determinate how human a robot is. It may come in the near future. However, all robotic researchers seem to agree that the robot would have to be able to show some social awareness and personality, and be capable of understanding and recognising people's speech and expressions.

But do we want robots to have more personality and to be able to take more initiative? Ultimately, to act more like us? Some may argue, yes. If intelligent vacuum cleaners were able to differentiate a sleeping human from objects, for example, at least one lady in South Korea wouldn't have had her hair "eaten" by her new domestic appliance.

But others argue that it is dangerous to give robots too much knowledge. And would it allow them to answer back? We are still at the beginning of research regarding the potential consequences this might have. Indeed, the scientific community is still debating whether a robot can ever have feelings or be self-conscious. Although AI has been able to perform certain tasks extremely skillfully, for example Alpha Go; the community is still a long way from developing an AI which closely resembles the human mind.

At present, robot companions are either focused on companionship or on task-execution. Jibo, for example, is a social robot that can talk, order food, remind you of things, or take pictures, while Roomba is an intelligent, but ultimately functional, vacuum cleaner.

Who is in charge?

But it's about striking the right balance, depending on the job at hand and the person it is working for. One most recent study showed that the more controlling and anxious about robots a person is, the more initiative they expect the robot to show and the more willing they are to delegate tasks to it. The research focused specifically on what level of initiative people preferred their robot companion to have when executing a cleaning task.

Participants could choose between manually turning on the cleaning robot themselves, having their robot companion turn on the cleaning robot remotely when instructed, or having the robot companion turn on the cleaning robot when it noticed that cleaning needed to be done. It was found that most people wanted their robot companion to execute the task without being asked.

This paradoxical result may be explained by the fact that people are now more used to technology – from computers and smartphones to smartwatches and intelligent home appliances – acting semi-autonomously. Smart companion robots are just the next step in the long evolution of our relationship with technology.

In the future, it is likely that we will see more domestic robot companions that can be customized to people's individual preferences. And we will be able to shop for them as we now shop for vacuum cleaners and phones. Ultimately, it seems, there will be a robot for everyone.

Artificial Intelligence Might Disrupt Science

In 2011, Artificial Intelligence (AI) came of age when IBM's Watson computer beat two human contestants to win Jeopardy. These were not any two contestants. Ken Jennings had won 74 times consecutively and Brad Rutter had pocketed the biggest pot in history - $3.25 million. In the battle between man and machine, Watson's win was historic. Jennings was sanguine about losing: "I, for one, welcome our new computer overlords."

The key to Watson's success is a technique called "deep learning", which has achieved astonishing results in several domains, most notably in understanding natural language. Research teams use it to teach computers to find meaning in vast amounts of text. Disrupting quiz shows is one thing, but can AI disrupt science?

Perhaps we should first ask, does science need disrupting? Yes. Access to reliable knowledge – the academic literature – is becoming a fundamental bottleneck for humanity. There are now over 50 million research papers and this is growing at a rate of over one million a year. Over 70,000 papers have been published on a single protein – the tumor suppressor p53. How can any academic keep up? And how can anyone outside of academia make sense of it all - the public, policy makers, business people, doctors or teachers? Well, most academics struggle and the public can't – most research is locked behind pay walls.

Ironically, in the age of the internet and unparalleled access to information, the most critical is out of bounds. Moreover, while we are clearly pretty good at producing knowledge, using this knowledge – that is separating the wheat from the chaff and integrating this together into something useful – is a big problem particularly in fields such as global sustainability.

That may be about to change. Here are five ways AI looks set to disrupt science.

1. Science mining: Iris. AI

The co-founders of this new AI start-up believe that if you could access and contextualize all of the world's published research you would solve a lot more problems. So they've set out to do that. "We want to democratize access to scientific knowledge. The first step is a science assistant leveraging AI and the crowd to help users map out and find relevant scientific knowledge," says Finnish co-founder Maria Ritola. The team has created an AI tool for innovators to do quick mappings of a research area, but in the long-term they want to build an AI scientist that can create a hypothesis based on existing publications, run experiments and simulations and even publish papers on the results. So they are not short of ambition.

They started with a simple tool to map out the science around a TED talk. Iris analyses the scripts of the talks using Natural Language Processing algorithms; mines open-access academic literature to find key papers related to the talk's content; then visualizes, really quite beautifully, the groups of related research papers. "Iris is a young AI. We call her our baby. She doesn't get everything right yet - she's at about 70% accuracy - so we're enlisting a community of AI Trainers to help her learn," Ritola explains. The TED tool is only the tiny first step, and in September the Iris team will launch the first commercial tool with a group of corporate R&D departments as pilot users.

2. Science mining: Semantic Scholar

This is genius. Or will be one day. Semantic Scholar is an academic search engine from Microsoft co-founder Paul Allen's Allen Institute for Artificial Intelligence. It too uses AI to search the academic literature and it is impressively fast. Still in beta, its interface is less elegant, more "academic utilitarian" than the sleek Iris. And it also throws up some oddities. A search for the landmark paper "A Safe Operating Space for Humanity", which appeared in *Nature* (along with *Science*, one of the two leading academic journals) in 2009, does not show up on the first page, nor does the follow up paper, which appeared in *Science* in 2015. The full-length version of the original paper, published in the more obscure journal *Ecology and Society*, is, however, the first entry. But, given the team only started in 2015, and they are first focusing on computer science papers, this is not bad going.

The team is adding new features. It can already trawl through a paper's reference list and work out which citation has been genuinely influential and which is just background.

3. From miner to scientist

A team at IBM has gone a step further than both Iris, AI and Semantic Scholar. They say their system can *do* science. That is, their AI can generate scientific hypotheses automatically by mining academic literature. Moreover, their algorithms, they say, can be used to make new scientific discoveries. Their goal is to combine text mining with visualization and analytics to identify facts and suggest hypotheses that are "new, interesting, testable and likely to be true", as the authors say in a 2014 research paper.

4. Science media: Science_Surveyor

Science journalists are the target for this bit of AI from a collaboration between Columbia and Stanford universities. Science Surveyor has been designed to help journalists assess the significance of a new piece of research. Where does the research fit into the bigger picture in the field? Is it really ground-breaking? Is it contested? Journalists need answers to all these questions on impossible deadlines and often with little expert knowledge in the area. This leads to churnalism, poor reporting and a readership that is none the wiser. Science Surveyor is an interesting experiment to move beyond this.

5. Open Access AI: Open AI

Sponsored by PayPal founders Elon Musk and Peter Thiel, among others, Open AI is a non-profit research company that aims to democratize AI. "It's really just trying to increase the probability that the future will be good," Musk said at the recent Recode Conference. Its goal "is to advance digital intelligence in the way that is most likely to benefit humanity as a whole, unconstrained by a need to generate financial return." This is a wildcard. No one is really sure what could come out of it, but if Musk is behind it, expect the unexpected – fast.

AI could be on the cusp of driving the next phase of the scientific revolution, yet this is less discussed than the bigger existential threat of AI. In many ways, AI innovations could simply help scientists to do their jobs more efficiently – thereby cutting the crippling time lag between science and society. For example, could machine learning algorithms delve deep into the previous five assessment reports of the Intergovernmental Panel on Climate Change and, based on research published since the last report, provide rudimentary conclusions of the sixth report?

One major hurdle to progress is the academic publishers. They have no financial incentives to grant AI initiatives access to the body of human knowledge, though Google Scholar has permission to trawl all text behind paywalls, so this may not be insurmountable.

Recently there was the launch of the Future Earth Media Lab to break down the barriers between science and society. Lab co-founder, and author of this piece, Owen Gaffney is a mentor for the Iris.AI project. Our mission is to seed, nurture and develop similar initiatives that can – project by project – nudge the technological revolution towards better outcomes for science and society, ultimately towards sustainable futures.

Are Killer Robots a Necessity?

New technology could lead humans to relinquish control over decisions to use lethal force. As artificial intelligence advances, the possibility that machines could independently select and fire on targets is fast approaching. Fully autonomous weapons, also known as "killer robots," are quickly moving from the realm of science fiction toward reality.

These weapons, which could operate on land, in the air or at sea, threaten to revolutionize armed conflict and law enforcement in alarming ways. Proponents say these killer robots are necessary because modern combat moves so quickly, and because having robots do the fighting would keep soldiers and police officers out of harm's way. But the threats to humanity would outweigh any military or law enforcement benefits.

Removing humans from the targeting decision would create a dangerous world. Machines would make life-and-death determinations outside of human control. The risk of disproportionate harm or erroneous targeting of civilians would increase. No person could be held responsible.

Given the moral, legal and accountability risks of fully autonomous weapons, preempting their development, production and use cannot wait. The best way to handle this threat is an international, legally binding ban on weapons that lack meaningful human control.

Preserving empathy and judgment

A report by Human Rights Watch and Harvard Law School states the belief that humans should dictate the selection and engagement of targets. The report was released in

April 2016 by Human Rights Watch and the Harvard Law School International Human Rights Clinic - two organizations that have been campaigning for a ban on fully autonomous weapons.

Retaining human control over weapons is a moral imperative. Because they possess empathy, people can feel the emotional weight of harming another individual. Their respect for human dignity can – and should – serve as a check on killing.

Robots, by contrast, lack real emotions, including compassion. In addition, inanimate machines could not truly understand the value of any human life they chose to take. Allowing them to determine when to use force would undermine human dignity.

Human control also promotes compliance with international law, which is designed to protect civilians and soldiers alike. For example, the laws of war prohibit disproportionate attacks in which expected civilian harm outweighs anticipated military advantage. Humans can apply their judgment, based on past experience and moral considerations, and make case-by-case determinations about proportionality.

It would be almost impossible, however, to replicate that judgment in fully autonomous weapons, and they could not be preprogrammed to handle all scenarios. As a result, these weapons would be unable to act as "reasonable commanders," the traditional legal standard for handling complex and unforeseeable situations.

In addition, the loss of human control would threaten a target's right not to be arbitrarily deprived of life. Upholding this fundamental human right is an obligation during law enforcement as well as military operations.

Judgment calls are required to assess the necessity of an attack, and humans are better positioned than machines to make them.

Promoting accountability

Keeping a human in the loop on decisions to use force further ensures that accountability for unlawful acts is possible. Under international criminal law, a human operator would in most cases escape liability for the harm caused by a weapon that acted independently. Unless he or she intentionally used a fully autonomous weapon to commit a crime, it would be unfair and legally problematic to hold the operator responsible for the actions of a robot that the operator could neither prevent nor punish.

There are additional obstacles to finding programmers and manufacturers of fully autonomous weapons liable under civil law, in which a victim files a lawsuit against an alleged wrongdoer. The United States, for example, establishes immunity for most weapons manufacturers. It also has high standards for proving a product was defective in a way that would make a manufacturer legally responsible. In any case, victims from other countries would likely lack the access and money to sue a foreign entity. The gap in accountability would weaken deterrence of unlawful acts and leave victims unsatisfied that someone was punished for their suffering.

An opportunity to seize

At a U.N. meeting in Geneva in April 94, countries recommended beginning formal discussions about "lethal autonomous weapons systems." The talks considered whether these systems should be restricted under the Convention on Conventional Weapons, a disarmament treaty that has regulated or banned several other types of weapons, including incendiary weapons and blinding lasers. It is crucial that the members agree to start a formal process on lethal autonomous weapons systems in 2017.

Disarmament law provides precedent for requiring human control over weapons. For example, the international community adopted the widely accepted treaties banning biological weapons, chemical weapons and landmines in large part because of humans' inability to exercise adequate control over their effects. Countries should now prohibit fully autonomous weapons, which would pose an equal or greater humanitarian risk.

While the process of creating international law is notoriously slow, countries can move quickly to address the threats of fully autonomous weapons. They should seize the opportunity presented by the review conference because the alternative is unacceptable: Allowing technology to outpace diplomacy would produce dire and unparalleled humanitarian consequences

Teach a Machine to Think

A Google search for Roger Federer, the Swiss tennis star, yields some 28,900,000 hits. International football star Lionel Messi even has as many as 61,300,000 entries. But there's one name that beats both of them hands down: searching for AlphaGo, the computer that defeated a master player of the strategy game Go in March 2016 returns no fewer than 313,000,000 hits. AlphaGo dominated the headlines in spring: machine triumphs over man. For some, AlphaGo's victory was the ultimate horror scenario, while others saw it as the breakthrough of artificial intelligence.

The master players

Joachim Buhmann, Professor for Computer Science and Head of the Institute for Machine Learning at ETH Zurich, offers a soberer assessment of the situation: "The Go player's algorithm has, of course, set a milestone in machine learning, but it's a milestone in a very limited, artificial field," he says. Since the early days of computer science as a scientific discipline, one of the challenges against which it has been relatively easy to measure progress has been strategy games. It started with simple games such as Nine Men's Morris and Draughts. In 1997, IBM's computer Deep Blue beat the reigning chess world champion Garry Kasparov. Soon thereafter, programmers set their sights on the considerably more complex game Go as the next potential milestone.

What is interesting, however, is not the fact that AlphaGo has now claimed victory, but rather how it did so: unlike Deep Blue, it didn't rely on sheer computing speed, but rather on enormous computing power "combined with a kind of clever learning," explains Buhmann. But he qualifies this by adding: "Successfully solving such game problems isn't the major breakthrough, because real intelligence is characterised by making a decision in the face of great uncertainty. And the game setting drastically reduces uncertainty." His research colleague holds a similar view: Thomas Hofmann is Co-Director of the new Center for Learning Systems, a joint endeavour between ETH and the Max Planck Society. In his words, "We want to build machines that succeed in the real world. Self-driving cars, for instance, are confronted with far more complex and consequential decisions."

Training in the sea of data

Nevertheless, the approach taken by the creators of AlphaGo to lead their computer to the championship is typical for many other areas of machine learning, as well. AlphaGo's designers first fed the machine with 150,000 matches that had been battled out by good players, and used an artificial neural network to identify typical patterns in these matches. In particular, the computer learned to predict which move a human player would make in a given position. The designers then optimized the neural network by repeatedly having it play against previous versions of its own games. In this way, through small but constant adjustments, the network gradually improved its chances of winning. "There are two ingredients that enable this type of learning," explains Hofmann. "You need a lot of data as learning material," he says, "and sufficient computing speed." Both are available today in many areas.

This dramatically changed the approach of developers in the field of artificial intelligence. Buhmann explains this based on the example of image recognition: previously, image experts had to tell the computer in detail which features it should use to categorize an image as a face, for example. "This meant that we had to rely on the knowledge of experts, and also that we had to describe vast amounts of rules in code," he recalls. Today, it is sufficient to write a meta-program that merely defines the basic principles of learning. The computer then learns by itself to tell, based on numerous sample images, which features depict a face. Thanks to Facebook, Instagram, etc., there is no shortage of learning material: "Today we can easily use millions of pictures or more as practice material," says Buhmann.

Computers as doctors

Buhmann specializes in image recognition in the medical field. As he explains, this field is precisely where the advantage of machine learning is clearly evident: "We used to try to ask doctors about their specialist knowledge and then implement it in detailed rules," he recalls, "but that endeavour ended in a terrific failure, because even good doctors often cannot provide clear explanations for their actions." Today, computer programs independently trawl through large volumes of image data for statistically relevant patterns. One specific area in which Buhmann and his colleagues use this type of method is cancer research, but the approach is also useful in studying neurological diseases such as schizophrenia, or neurodegenerative diseases such as dementia or Parkinson's disease.

They have, for instance, developed a program that helps pathologists to assess the likely development of a certain form of kidney cancer more accurately. The process involves obtaining patient biopsies and preparing histological sections, using certain dyes to make relevant features visible. The sections are digitized and analyzed using machine image analysis methods in order, for example, to count the cancer cells that are in the process of dividing and that were made visible by the staining. The computer then combines such counts with additional data to develop prognoses for specific patient groups. In another project, computers were used to analyze magnetic resonance images of the brains of schizophrenia patients. The image analysis yielded three groups of patients with significantly different activity patterns in the brain. "We learned that there are different kinds of schizophrenia," explains Buhmann, adding: "Now it's up to pharmacists and doctors to find the right treatment for each patient type." It is quite possible that automated analyses of brain images will help with this, too.

Language and meaning

Image recognition is to Buhmann what language is to his research colleague Hofmann. "Speech recognition as a branch of artificial intelligence is in particular demand when it comes to human-machine interaction," explains Hofmann. He hopes that he will one day no longer have to tediously enter his desired destination via a keyboard to let a self-driving car know where he wants to go, but can instead spontaneously give it oral instructions. Hofmann is convinced that it won't be long now until this happens: "Today, we can approach the problem of getting machines to understand text in a completely different way from how we could before."

Here, too, big data supplies the material the machines use to practice understanding texts. The web is a vast treasure trove of language, a gigantic training ground that helps machines filter out statistical regularities that show them relationships between words. "And it does a much better job of it than we could ever have done with abstract linguistic or phonetic rules," says Hofmann. This kind of method can also be used to optimize translation programs or search engines. Hofmann and his team are developing a program that uses all Wikipedia entries (there are more than 5 million English-language articles) as a basis for learning to link texts and words in a way that makes sense. The links and cross references to other articles, which Wikipedia authors currently still create manually, will in future be added by a computer – faster and more comprehensively than any author would be able to manage. "It starts with the fundamental meanings of words. But then our goal is to get our programs to understand the meaning of complete sentences and, ultimately, entire discourses," says Hofmann.

On an equal footing with machines

Pie in the sky? Only partly. Translation programs have already made tremendous progress in recent years. Search engines are constantly improving and computer programs are now authoring sport updates. Hofmann himself was involved in founding a company called Recommind in the US. Their programs analyze and sort texts with a view to their legally relevant content. "We automate document review, which used to take lawyers endless hours," he explains. Today this company employs 300 staff worldwide and is the market leader in its field.

Recommind is just one example of how new technologies will change even jobs in highly qualified professions. Hofmann is convinced that there are only a few occupations that will not feel the impact of this technological change. "To date, machines have taken over repetitive, mechanical jobs. In future, they will also take intelligent decisions," he says. Buhmann, too, is confident, claiming that "the new intelligent technologies will in future supplement or even replace activities performed by well-trained specialists." For instance, the new possibilities in image analysis will no doubt massively change the work of pathologists. As Buhmann points out: "We will need far fewer pathologists in future – but that means doctors could spend more time on psychological care for the sick." His colleague Hofmann adds: "In terms of technology, everything is possible. It's a question of society's willingness to find creative solutions for dealing with this technological change."

Dawn of Creative Artificial Intelligence

Data scientists are creating robo-artists that use machine learning to create artworks and even develop an artistic style. Something went very wrong with one of Google's neural networks. It was designed for a simple task: identify dogs in photos. But a curious developer reversed the algorithm and it began to hallucinate dogs where there were none before. The psychedelic images resembled those of Salvador Dali, and echoed across the internet with the short hand "Deep Dream".

Within a few months of this discovery, an academic paper repeated the same magic feat for famous painters. Data scientists built a set of robo-artists out of digital neuron clusters called recurrent neural networks. They used machine learning and artificial intelligence to reverse engineer visual art resembling Picasso's dancing lines, Van Gogh's hypnotic brush strokes and Edvard Munch's emotional impact. We have taught robots how to make art by teaching them what makes an artistic style. And so "Deep Style" was born.

What should we make of our creative automatons? In human affairs, children of successful lawyers and accountants often have the freedom to become creators, liberated from monetary constraints and able to dance, paint and make music. This time, our software progeny are transcending their humble beginnings. They just might become humanity's greatest artists, amplifying and robotizing creativity.

The computer revolution has catalyzed tremendous automation: in physical labor in places like factories; and increasingly now in intellectual labor, from legal discovery to roboadvisors. As the Marc Andreessen saying goes,

"Software is eating the world". A recent McKinsey study projects that 45% of all office work will be automated in the near future. Software processes our paperwork, searches for results, takes payments, directs cars, and talks with other systems to create lattices of efficiency. But our programs to date have been deeply analytical, following prescribed top-down rules to implement productivity tasks.

That left-brained set of rigid algorithms is about to meet its right-brained counterpart. The key is that this new sort of software isn't replicating a set of rules to distort an image per human design. Rather, it is using sophisticated math to process visual information, extract unique patterns, and recursively learn what makes any particular artistic style unique. Then it can take off from there. Think of it as statistical intuition, not unlike our own instincts and gut impulses. Mobile apps like Dreamscope (free, amazing, on iOS/Android) allow a user to apply this machine-learned creativity to a photo on command. Dreamscope has indexed dozens of creative algorithms—a robot for each painter—and enables a user to "seed" their own machine artist. How long until every creative human endeavor has been patterned in this way?

Already, we find machine learning applications in the visual arts, music and writing. The programs are young and often spit out creations that seem somehow wrong, though we cannot put a finger on why. These machine artbots are from the wrong side of the Uncanny Valley – a category of things that attempt to mimic humanity but in their artifice create unease.

And yet, we have never been closer to a room of monkeys typing out the collected works of Shakespeare. Just ask a robot that has ingested all of Shakespeare's works and is trained to generate soulful prose on command, ad infinitum. Or turn on machine-Bach, mathematically generating emotional sound vibrations that, someday, may be indistinguishable from the real thing. The following texts are neural network generated samples based on Shakespeare, which can be created ad infinitum.

KING LEAR:

One loyal of my love, the wedding-body touchest thee: I pray,

Henceforwards, and submiss the truth! though my throne

Lives as mock'd my pardon with some untold

Attore sack lop and shrum' them up:

But be preserved with spirits, so brimfibed again!

My voices were so early, I was enough.

MACBETH:

Then let him withdraw them debour to branch ere any

day, but to prevail'd be penny of a merry tongue

Which the exploits of fools look with their veins.

Beware, artists. Automation will impact not only the analytical industries, but also those that require creativity, originality and intuition—domains that were once believed to be uniquely human. If you are an artist, musician, or writer, artificial intelligence is about to present challenges and opportunities that rivaled the ones posed to painters by the invention of photography in the 1800s. What now seems like a crude hollow reproduction of a mystical human endeavor could eventually be responsible for the bulk of all art, initiated by humans but outsourced to machines.

There are many objections to the idea that true art can even be made by software. Isn't the human always the root of the process? Isn't the artist's impulse to create profoundly human? Isn't the point of art to, in some way, symbolize and instantiate the unique point of view of the human artist in order to evoke a uniquely human response in the viewer or listener? Aren't our cultural values—a result of the arbitrary and arduous evolution of a mammalian body—the only lens capable of authoring and appreciating art, as such? So what will be the message or set of values implicit in machine-generated art? These questions are fair, but in my opinion only partially relevant.

As the shift toward the machine continues, there will be increasingly less space for human execution of what qualified for creative endeavors in the past. Instead of composing music, we will create randomization algorithms that combine software-composers on the fly, reacting to our quantified moods and surroundings. Instead of learning to paint, aspiring artists will be better served to learn how to code programs that render creative outcomes in simulated virtual reality environments.

The raw materials for this revolution are in place. Wearable sensors will make it possible to create an essentially infinite data set of the images, sounds and text that humans exchange every day. Google Photos and other cognitive computing tools are processing millions of such inputs daily. Our culture can increasingly be mapped, studied and statistically modeled. Hard rules about aesthetics are not necessary when we can just point our learning machines to the recorded history of what humans believe is beautiful and meaningful. The Golden Ratio is timeless.

What will be the meaning of such "art"? Critics of the future will wrestle with such questions.

We can also simulate evolution and reward the most creative software with fitness and something resembling life itself. In 2013, engineers at Cornell Creative Machines Lab used evolutionary programming to create 3D cubes that learned how to walk: the randomized critters that ambled fastest were allowed digital offspring that moved faster with each generation.

Our robo-artist could be motivated by a different outcome – to move the human spirit – using the vast data generated by human activity both as inputs and to determine the success and impact of their new creations. Yes, humans will set many of the creative programs into motion, but the ultimate outcome will be the product of machines. We will be the builders, accountants, and lawyers—our digital children will dance, paint and sing.

The Booming Market of Robots

The robotics industry is experiencing an investment boom, as the technology develops. What could it mean for humans?

In warehouses, hospitals, and retail stores, and on city streets, industrial parks and the footpaths of college campuses, the first representatives of this new invading force are starting to become apparent.

"The robots are among us," says Steve Jurvetson, a Silicon Valley investor and a director at Elon Musk's Tesla and SpaceX companies, which have relied heavily on robotics. A multitude of machines will follow, he says: "A lot of people are going to come in contact with robots in the next two to five years."

The arrival of the robots — and their potentially devastating effect on human employment — has been widely predicted. Now, the machines are starting to roll or walk out of the labs. In the process, they are about to tip off a financing boom as robotics — and artificial intelligence — becomes one of the hottest new markets in tech.

After growing at a compound rate of 17 per cent a year, the robot market will be worth $135bn by 2019, according to IDC, a tech research firm. A boom is taking place in Asia which is in the early stages of retooling its manufacturing sector - Japan and China accounting for 69% of all robot spending.

Although the amount of money flowing into a new robotics industry is still at a relatively early stage, all the lead indicators of the innovation economy are pointing up. Patent filings covering robotics technology — one sign of the expected impact — have soared. According to IFI Claims, a patent research company, annual filings have tripled over the past decade. China alone accounted for 35 per cent of robot-related patent filings last year — more than double its nearest rival Japan.

In another sign of the expected boom, venture capital investments more than doubled last year to $587m, according to research firm CB Insights.

Other investors are also piling in, says Manish Kothari of SRI International, a Silicon Valley research and development lab that has spun off robot companies. From private equity investors looking to build portfolios of robot investments, to new "incubators" such as Playground, started by former Google robotics chief Andy Rubin, the investment options have been proliferating rapidly.

But in many cases, the amounts being invested still seem disarmingly modest. Like other disruptive technologies, the seeds of this revolution can be seen in start-ups that operate on a shoestring but have grandiose aims.

They include companies such as Dispatch, a Silicon Valley company that is testing an autonomous delivery vehicle — a smart-box on wheels — on two college campuses in the US. The start-up has raised only $2m but is riding the wave in collapsing costs of sensors and advances in artificial intelligence that are making autonomous machines a reality.

"There is an exponential pace of improvement in hardware and machine learning algorithms," says co-founder Uriah Baalke. "The computational power required has gone down a lot." The result is a new class of machines that can operate by themselves in human space, the advance guard of a new robot industry.

Until now, most robots have taken the form of expensive, high-precision industrial machines. Usually found operating in protective cages on automobile assembly lines, they have carried out preprogrammed tasks, with no need or scope to adapt to changing conditions.

The cheaper, flexible machines that are emerging are designed to be more adaptive. From driverless cars and drones to the "cobots" that work alongside humans in industrial settings, they try to sense and adapt to their surroundings. Like Tug (a robot that moves supplies around hospitals), Savioke (which handles deliveries to hotel rooms) and Locus Robotics (which operates in warehouses), they are moving into the service industries. In industrial settings — still the main venue for robot investment — they are moving out of the cages and into a far wider range of roles.

Like the arrival of PCs, the new era promises to take the technology into many more areas of working life. "The traditional industrial robots are mainframes — what we're doing are PCs," says Scott Eckert, chief executive of Rethink Robotics, a US company whose robots help with packing or tend machines. Rethink says that the all-in cost of its Sawyer robotic arm amounts to about $1 an hour, a price at which many of the jobs that have been beyond the reach of automation could be affected.

The technology advances behind this wave of innovation have come together remarkably quickly. Funding over the past five years by Darpa, the research arm of the US defence department, has brought breakthroughs in mechanical areas such as robotic limbs, says SRI International's Kothari.

But the biggest advances have come in software. Improvements in computer vision, for instance, have made possible many companies like Dispatch, whose machines rely on being able to "see" the world around them, says Chris Dixon, a partner at venture capital firm Andreessen Horowitz.

Machine learning algorithms, which are designed to adapt through an endless process of trial and error, play the biggest part in teaching robots how to navigate a world beyond the normal rules-based systems that computers are designed to handle.

"You won't have to programmatically tell it what to do; it will figure it out," says Vinod Khosla, a venture capitalist who has backed robot companies in markets including agriculture and healthcare. "Today, it's really dumb intelligence — but that will change quickly."

When it comes to designing the machines for this emerging industry, most robot entrepreneurs and investors are following a similar formula.

One element is to build low-cost machines that tackle specific tasks, rather than attempt to create general-purpose machines — let alone fully humanoid robots — that try to take on too much.

The goal is to build "single-purpose robots that do one thing very well", says Dmitry Grishin, a Russian who recently raised a $100m fund to invest in robots and other hardware. If they succeed, these machines quickly lose their status as "robots" and become more part of the fabric of everyday life, he says — like automated vacuum cleaners or cash machines.

Another design feature of many of the early robots is to operate alongside people, making humans more productive rather than replacing them altogether. Many of these robots, for instance, hand over decision-making to a human operator when they encounter situations they cannot understand or navigate.

"The truth is, anyone who works in robotics knows the limitations of what they're working with, and they're pretty extensive," says Kothari. Robot companies also want to keep "the human in the loop" because they believe it will make their machines more socially acceptable and less threatening, he says. Most people operating in the robot industry say humans will have an important role to play in directing the machines for decades to come.

That does not change the long-term threat to jobs, however. "There isn't a single mechanical or physical thing a human will be able to do better than a robot," says Tesla and SpaceX's Jurvetson.

Another feature the robot makers are counting on is to be able to use the learning capabilities of their initial products to achieve rapid improvements and gain an advantage over rivals that are slower to get their machines into the market.

"Once you ship the device, you can apply more and more intelligence and machine learning," says Grishin, the Russian robot investor. The trick, he says, will be to find a task that the relatively dumb machines are able to handle, then use knowledge gained in the field to rapidly add to their capabilities and usefulness. "First put them in consumers' hands, then learn from their behaviour."

This is the secret weapon that all robot companies rely on. "Everything gets better over time," says Jurvetson. "This is happening in almost every hardware product: they are becoming minimal vessels for software."

This technological shift has set traditional robotics leaders in Japan and Germany against nascent industries in countries such as the US and China.

"Right now, the US is definitely the leader" when it comes to software, says Grishin. He adds, however, that the hardware manufacturing expertise of China makes that country a contender, particularly since robotics has become a national priority. As a result, the rise of a new robot industry is about to trigger a global race for leadership.

Robots Need Ability to Say No

Robots should learn when it's appropriate to follow human commands and when it's not.

Should you always do what other people tell you to do? Clearly not. Everyone knows that. So, should future robots always obey our commands? At first glance, you might think they should, simply because they are machines and that's what they are designed to do. But then think of all the times you would not mindlessly carry out others' instructions – and put robots into those situations.

Just consider:

- An elder-care robot tasked by a forgetful owner to wash the "dirty clothes," even though the clothes had just come out of the washer
- A preschooler who orders the daycare robot to throw a ball out the window
- A student commanding her robot tutor to do all the homework instead doing it herself
- A household robot instructed by its busy and distracted owner to run the garbage disposal even though spoons and knives are stuck in it.

There are plenty of benign cases where robots receive commands that ideally should not be carried out because they lead to unwanted outcomes. But not all cases will be that innocuous, even if their commands initially appear to be.

Consider a robot car instructed to back up while the dog is sleeping in the driveway behind it, or a kitchen aid robot instructed to lift a knife and walk forward when positioned behind a human chef. The commands are simple, but the outcomes are significantly worse.

How can we humans avoid such harmful results of robot obedience? If driving around the dog were not possible, the car would have to refuse to drive at all. And similarly, if avoiding stabbing the chef were not possible, the robot would have to either stop walking forward or not pick up the knife in the first place.

In either case, it is essential for both autonomous machines to detect the potential harm their actions could cause and to react to it by either attempting to avoid it, or if harm cannot be avoided, by refusing to carry out the human instruction. How do we teach robots when it's OK to say no?

How Can Robots Predict?

Scientists have started to develop robotic controls that make simple inferences based on human commands. These will determine whether the robot should carry them out as instructed or reject them because they violate an ethical principle the robot is programmed to obey.

Telling robots how and when – and why – to disobey is far easier said than done. Figuring out what harm or problems might result from an action is not simply a matter of looking at direct outcomes. A ball thrown out a window could end up in the yard, with no harm done. But the ball could end up on a busy street, never to be seen again, or even causing a driver to swerve and crash. Context makes all the difference.

It is difficult for today's robots to determine when it is okay to throw a ball – such as to a child playing catch – and when it's not – such as out the window or in the garbage. Even harder is if the child is trying to trick the robot, pretending to play a ball game but then ducking, letting the ball disappear through the open window.

Explaining morality and law to robots

Understanding those dangers involves a significant amount of background knowledge (including the prospect that playing ball in front of an open window could send the ball through the window). It requires the robot not only to consider action outcomes by themselves, but also to contemplate the intentions of the humans giving the instructions.

To handle these complications of human instructions – benevolent or not – robots need to be able to explicitly reason through consequences of actions and compare outcomes to established social and moral principles that prescribe what is and is not desirable or legal. As seen above, our robot has a general rule that says, "If you are instructed to perform an action and it is possible that performing the action could cause harm, then you are allowed to not perform it." Making the relationship between obligations and permissions explicit allows the robot to reason through the possible consequences of an instruction and whether they are acceptable.

In general, robots should never perform illegal actions, nor should they perform legal actions that are not desirable. Hence, they will need representations of laws, moral norms and even etiquette to facilitate determining whether the outcomes of an instructed action, or even the action itself, might be in violation of those principles.

While our programs are still a long way from what we will need to allow robots to handle the examples above, our current system already proves an essential point: robots must be able to disobey to obey.

Can Artificial Intelligence be Ethical?

AlphaGo, a computer program specially designed to play the game Go, caused shockwaves among aficionados when it defeated Lee Sidol, one of the world's top-ranked professional players, winning a five-game tournament by a score of 4-1.

Why, you may ask, is that news? Twenty years have passed since the IBM computer Deep Blue defeated world chess champion Garry Kasparov, and we all know computers have improved since then. But Deep Blue won through sheer computing power, using its ability to calculate the outcomes of more moves to a deeper level than even a world champion can. Go is played on a far larger board (19 by 19 squares, compared to 8x8 for chess) and has more possible moves than there are atoms in the universe, so raw computing power was unlikely to beat a human with a strong intuitive sense of the best moves.

Instead, AlphaGo was designed to win by playing a huge number of games against other programs and adopting the strategies that proved successful. You could say that AlphaGo evolved to be the best Go player in the world, achieving in only two years what natural selection took millions of years to accomplish.

Eric Schmidt, executive chairman of Google's parent company, the owner of AlphaGo, is enthusiastic about what artificial intelligence (AI) means for humanity. Speaking before the match between Lee and AlphaGo, he said that humanity would be the winner, whatever the outcome, because advances in AI will make every human being smarter, more capable, and "just better human beings."

Will it? Around the same time as AlphaGo's triumph, Microsoft's "chatbot" – software named Taylor that was designed to respond to messages from people aged 18-24 – was having a chastening experience. "Tay" as she called herself, was supposed to be able to learn from the messages she received and gradually improve her ability to conduct engaging conversations. Unfortunately, within 24 hours, people were teaching Tay racist and sexist ideas. When she started saying positive things about Hitler, Microsoft turned her off and deleted her most offensive messages.

I do not know whether the people who turned Tay into a racist were themselves racists, or just thought it would be fun to undermine Microsoft's new toy. Either way, the juxtaposition of AlphaGo's victory and Taylor's defeat serves as a warning. It is one thing to unleash AI in the context of a game with specific rules and a clear goal; it is something very different to release AI into the real world, where the unpredictability of the environment may reveal a software error that has disastrous consequences.

Nick Bostrom, the director of the Future of Humanity Institute at Oxford University, argues in his book *Superintelligence* that it will not always be as easy to turn off an intelligent machine as it was to turn off Tay. He defines superintelligence as an intellect that is "smarter than the best human brains in practically every field, including scientific creativity, general wisdom, and social skills." Such a system may be able to outsmart our attempts to turn it off.

Some doubt that superintelligence will ever be achieved. Bostrom, together with Vincent Müller, asked AI experts to indicate dates corresponding to when there is a one in two chance of machines achieving human-level intelligence and when there is a nine in ten chances. The median estimates for the one in two chances were in the 2040-2050 range, and 2075 for the nine-in-ten chance. Most experts expected that AI would achieve superintelligence within 30 years of achieving human- level intelligence.

We should not take these estimates too seriously. The overall response rate was only 31%, and researchers working in AI have an incentive to boost the importance of their field by trumpeting its potential to produce momentous results.

The prospect of AI achieving superintelligence may seem too distant to worry about, especially given more pressing problems. But there is a case to be made for starting to think about how we can design AI to consider the interests of humans, and indeed of all sentient beings (including machines, if they are also conscious beings with interests of their own).

With driverless cars already on California roads, it is not too soon to ask whether we can program a machine to act ethically. As such cars improve, they will save lives, because they will make fewer mistakes than human drivers do. Sometimes, however, they will face a choice *between* lives. Should they be programmed to swerve to avoid hitting a child running across the road, even if that will put their passengers at risk? What about swerving to avoid a dog? What if the only risk is damage to the car itself, not to the passengers?

Perhaps there will be lessons to learn as such discussions about driverless cars get started. But driverless cars are not super intelligent beings. Teaching ethics to a machine that is more intelligent than we are, in a wide range of fields, is a far more daunting task.

Bostrom begins *Superintelligence* with a fable about sparrows who think it would be great to train an owl to help them build their nests and care for their young. So, they set out to find an owl egg. One sparrow objects that they should first think about how to tame the owl; but the others are impatient to get the exciting new project underway. They will take on the challenge of training the owl (for example, not to eat sparrows) when they have successfully raised one.

If we want to make an owl that is wise, and not only intelligent, let's not be like those impatient sparrows.

Equip Workforces for Future Jobs

The robots are coming and are taking our jobs. Or are they? The media and the blogosphere have been buzzing lately about the impact of artificial intelligence and robotics on our lives. In particular, the debate on the impact of automation on employment has amplified concerns about the loss of jobs in advanced economies. And accelerating technological change points the spotlight on questions like: Do workers, blue and white collar alike, possess the right skills for a changing labor market? Are they prepared for the employment shocks that come with the so-called "fourth industrial revolution"? What skills strategy should countries adopt to equip their workforces for the 21st century?

A good start to answering these questions is to define the concept of "skills" and understand skills formation. A worker's skill set has three components: cognitive skills like basic numeracy and literacy (including digital literacy), as well as advanced problem-solving and creative and critical thinking skills; social and behavioral skills like conscientiousness, grit, and openness to experience; and job or occupation-specific technical skills like those required to work as an engineer or electrician. Cognitive and social and emotional skill formation starts from very early childhood, with many social and behavioral skills remain malleable throughout adult life. But the window for building cognitive skills closes with late adolescence. This does not mean we stop learning new things, but our brains are hardwired at certain point in development. In contrast, technical skills are acquired later from adolescence throughout adulthood and in vocational schools, universities, and on the job. They depreciate, at an increasingly fast rate given technological change, and require constant upgrading.

Cognitive and social and behavioral skills are a priority

Skills strategies should prioritize the formation of cognitive and social and behavioral skills in early childhood and school. Good cognitive and social and behavioral skills are necessary for gaining and improving technical skills throughout life. They make workers more resilient to technology-driven labor market shocks like automation. Skilled workers have even begun replacing robots in modern car production because they are more adaptable and flexible. On the contrary, adults with poor literacy and numeracy skills have difficulty learning and updating the technical skills needed to compete in the modern job market. All over the world, including the United States and Europe, jobs are shifting from routine tasks, which are prone to automation, towards interactive tasks, which require advanced cognitive and social and behavioral skills.

Yet many youth and adults in advanced and emerging economies have considerable basic cognitive skill deficits. Take the European Union (EU) for instance. In a majority of EU countries, a fifth or more of 15 year-olds scored below functional literacy and numeracy in the mathematics and reading tests of the 2012 Program for International Student Assessment (PISA). Poor performers often come from socially disadvantaged backgrounds, suggesting that education does not always present an opportunity for social mobility. This is bad economic and social news for countries with declining and aging populations.

What does such a skills strategy mean for 21st century education policy? First, governments should prioritize universal basic cognitive skills like reading and mathematics, emphasizing quality education for children from disadvantaged background who are disproportionately represented among poor performers in PISA. This starts with expanding quality early childhood development and education interventions to help stimulate brain development during the all-important early years. Strengthening social and behavioral skills in school, for example through "growth mindset" interventions, can be an effective tool to make up for disadvantage by boosting students' confidence and goal orientation.

Second, governments can take advantage of the opportunity provided by longer years of schooling. With students "captive" in kindergarten and school during the critical period of their skills formation, innovation in classroom and teaching practices, such as in Colombia's Escuela Nueva model now adapted in PISA 2012's success story Vietnam, can help foster advanced skills like problem-solving, critical thinking and team work.

Third, the answer to the fast depreciation of technical skills is, beyond ensuring faster brains, a faster adaptation of technical education and training content through greater partnership between firms and universities and vocational schools. Technology and big data can help capture and better understand the evolution of occupations and technical skills needs in real time: Just let the robots work for us.

The Look of the Future Home

If the third industrial revolution was about using electronics and information technology to change economic systems and the way we live, the fourth will be characterized by disruptions stemming from a merger of the digital and physical worlds.

Much of the focus at the World Economic Forum's Annual Meeting in Davos has been on how the breakthroughs we are seeing today in artificial intelligence, robotics, the Internet of Things, biotechnology, and other fields are disrupting industry, and ushering changes in systems of production, management, and governance. But the changes and disruptions we are seeing are about so much more than economic systems and industry. AI and robotics are transforming the way we live.

Consider our homes. The home has developed from pre-industrial buildings that provided shelter and insulation, to industrial-era buildings that added water and waste disposal along with energy systems, to more recently adding simple sensors and networking during the electronic era with the third industrial revolution.

What we are seeing now with the emergence of the fourth industrial revolution is the development of cognitive architecture, which enables our living spaces to be tailored for personal and family preferences. This is set to have a profound effect on our quality of life.

The home will become a natural, intuitive, extension of you. Rather than the occupant adapting to the home, we've entered an exciting new phase where the home works for those who live inside it.

Development of AI, robotics, and other advanced technologies for applications within the living space has been underway for some time, but are gaining increased attention. Facebook founder Mark Zuckerberg, for instance, recently highlighted the potential of AI within the home and has set himself the challenge of creating a personal, voice-activated, intelligent home control and monitoring system. Meanwhile, companies such as Nest are creating connected products that recognize homeowners' preferences and adjust settings like temperature automatically or via an app.

In the same way that primary energy use in the home shifted from lighting to more complex devices and appliances, Internet traffic is following a similar pattern. Professor Klaus Schwab's report on the Fourth Industrial Revolution predicts that the tipping point will be when over 50% of internet traffic delivered to homes is for appliances and devices as opposed to entertainment and communication, and that we can expect this tipping point to have occurred by 2025.

The social benefits stand to be immense. At LIXIL, a Japan headquartered building and housing products and materials company, we are very close to the issues surrounding Japan's rapidly aging society. Better connectivity in the home and caregiving technologies are some of the most effective ways to enable aging in place and maintain quality of life.

Sensors can already detect movement throughout the home and body temperature, but the introduction of AI and machine learning will make it possible to assess with increasing accuracy when patterns of activity may be abnormal, allowing remote care providers or family to be alerted. Toilets, for example, will be able to monitor the body's health, providing an early warning on potential medical issues and when a trip to the doctor may be necessary.

In similar ways, increased connectivity and the use of sensors will provide enhanced security for those with young children. Parents who are out at work would have far greater peace of mind being able to know when their children safely returned home, while exterior sensors on the house also make it possible to make the home more secure. The list of potential applications is endless.

The accrual of incremental improvements in the home also has the potential to bring enormous benefits to global issues such as water conservation and climate change. The U.S. Energy Information Administration reported that 41% of total U.S. energy consumption in 2014 was consumed in residential and commercial buildings.

Connected thermostats coupled with sensors on windows and housing exteriors, for example, will make it possible to place automatic controls over temperature to help save energy, while more efficient toilets and plumbing systems are already enabling significant savings in water use.

There are of course significant challenges. Increased connectivity in today's world is, not surprisingly, bringing with it increased concerns around privacy and vulnerability to cybercrimes. Do we really want our homes, where we spend our private time, to be connected and "always on"? In addition, how do we ensure that new solutions are scalable and can benefit all including those in emerging markets, instead of just the privileged few?

For the "smart" connected home to realize its potential, these are all critical issues that need to be addressed. At LIXIL, we stand at a unique intersection of the movement to integrate the physical and digital worlds. We aim to bring together people, ideas, and technologies in unexpected ways to enhance the lives of people and society. Over the coming years, developments we are now seeing with the fourth industrial revolution will dramatically impact how these enhancements to our quality of life and society occur, including in the home.

Disruptive Technologies to Reshape Business

Regardless of your industry, the marketplace is continually evolving. The reason, increasingly, is the evolution of disruptive technology.

Disruptive technologies are enhanced or new technological innovations that essentially displace conventional and established technology, rendering it obsolete. They can create opportunities for new products, new markets, and new ways of conducting business.

Business models will again change as businesses adapt. The enhancement of current technology and the development of new technological innovations will undeniably transform how new businesses are established, and how existing businesses compete. For small and medium-sized firms, technology will also enable significant leaps forward in terms of innovation, efficiency, and competitiveness.

Adapting quickly will be essential; so here's the top six that businesses need to prepare for.

- **Social Robotics**

Robots, no longer restricted to the factory floor, are increasingly being designed to interact directly with humans. This not only means that – depending on your industry – certain robots may be about to enter your product ranges, but also that a robot may be interacting with customers on your behalf like the emotion-sensing Pepper robot, which according to TTG Asia has been "hired" to work on cruise ships.

Robots have previously been fixed to the factory floor, but they are set to become much smaller, more collaborative, and more affordable. Robotics are also making it possible for small companies to expand, such as Skyline Windows in New York, which rely on robots for installation of its windows.

• Artificial intelligence and smart services

True artificial intelligence - that which is so similar to human intelligence as to be indistinguishable - is difficult to develop, but that doesn't mean we won't be running into intelligent machines and smart services as the years progress.

Consider the applications of a machine or service that could learn about your customers, going beyond website analytics to truly understanding their day-to-day behaviour. Or IBM's Watson which uses natural language processing to enable partnerships between people and computers. The same technology could help you with business concerns from staffing to strategy. While we're not quite there yet, consider how smart services like Apple's Siri are already becoming more ubiquitous as smart phone ownership increases.

• Virtual reality

Originally considered a gaming technology, virtual reality is becoming more mainstream, and the applications for businesses and consumers are plentiful.

Consider how you could apply a completely immersive environment in your business, and how this might change the competitive landscape in your industry. For instance, in 2015 Volvo offered virtual reality test drives using Google's mass-produced virtual reality technology,

Cardboard. The technology has also been used to provide tours, make events more immersive, and even for training purposes.

- **3D printing**

Just as virtual reality offers us the ability to bring our thoughts into "reality" for consumers or colleagues, 3D printing offers us the chance to do this with physical reality. 3D printing lets us bring imagination into the physical world, whether we're showcasing prototype products to investors or custom-making products for consumers. 3D modelling and 3D printing are gradually changing consumer markets and have been used to create a range of products including musical instruments, medical equipment, artificial organs, and manufactured car parts.

- **The internet of things**

We already know that everyone is connected, but what about everything? This is the reality that the "internet of things" will bring. From small changes - your car communicating with your office to switch on the air conditioning, computer, and coffee machine moments before you arrive - to larger changes like your global offices being truly connected, beyond what is already offered by cloud computing to consumer applications.

Innovations like the Nest Protect allows users to "hush" a smoke alarm from a smartphone, essentially allowing the smoke alarm to speak to smart devices. The treatment of security and privacy concerns will determine the speed with which the internet of things rolls out.

- **Mobile and wearable technology**

Smartphone ownership is at an all-time high, bringing opportunities for businesses to take their operations truly mobile – and to contact consumers in new ways.

Developments in Near Field Communication technology allow us to know where consumers are (with their permission of course) and mean we could potentially send them relevant promotions based on their location, or remember their preferences for a whole new take on customer loyalty.

Consumers are also increasingly taking up wearable technology such as smart watches, pedometers, and ear pieces. This wearable technology, working with data on patterns and behaviour, could not only empower consumer interactions but make for more efficient, productive, and happier employees. Smart clothes are potentially the future of wearables - OMsignal already offers a line of smart shirts, and soon a sports bra which tracks biometric fitness data.

Disruptive technologies will significantly influence business models over the next few decades. A recent report from the Economist Intelligence Unit and Ricoh stated that "businesses will have nowhere to hide from the disrupting yet energising effects of technology change". The report suggests it's no longer viable to implement new technological innovations simply for short-term efficiency gains; instead technology disruption necessitates the implementation of new changes over time, for longer-term efficiency gains.

Technological Change - a New Global Economy?

Robot revolution! New technologies will dramatically change the nature of work across industries. Our lives are being shaken to their very core by technological change, with the Fourth Industrial Revolution transforming economies as never before.

The unprecedented speed of change, as well as the breadth and the depth of many radical changes unleashed by new digital, robotic and 3D technologies, is having major impacts on what we produce and do, how and where we do it and indeed how we earn a living. And while the transformation will proceed differently in advanced and developing parts of the world, no country or market will be spared from the tidal wave of change.

To appreciate the changes at hand, two interrelated aspects of the economy are particularly illustrative: growth and productivity on one hand, and employment on the other.

As the World Economic Forum highlights annually in its *Global Competitiveness Report*, productivity is the most important determinant of long-term growth. Yet productivity growth has stagnated around the world, particularly since the great recession, putting into question our ability to provide rising living standards for the world's citizens. While arguments abound as to what has been driving the productivity slowdown, an important question is how the Fourth Industrial Revolution will drive it in the years to come.

In theory, the application of new technologies to existing problems should improve efficiency and thus productivity.

Technological innovations tend to raise labour productivity by allowing the existing workforce to do more with less, by replacing existing workers with technology (with an obvious downside, as I will come to later), and they also usher in new products and processes that open new sources of growth.

Techno doom, or techno utopia?

Yet there is much debate on the likely size of the impact. On one hand, experts such as Robert Gordon of Northwestern University believe that the most important contributions of the digital revolution have already been made, and that the productivity impact of the current technological revolution is almost over. That would be worrisome indeed, particularly given the present slowdown.

On the other hand, "techno optimists" such as Eric Schmidt, the chairman of Google, believe that the world has reached an inflection point and will soon be experiencing faster growth and a major surge in productivity.

Perhaps there are such divergent views because the impact of technology is so difficult to measure. Even back in 1987, the Nobel-winning economist Robert Slow noted: "You can see the computer age everywhere but in the productivity statistics."

The trouble with Airbnb and GDP

Regardless of the precise effect on traditional measures of productivity and growth, inadequate measurement is an issue. The Ubers and Airbnbs of the world are clearly providing efficiency and productivity gains. Yet many of the benefits of these new activities are not accounted for in the calculation of GDP, in the same way that private housework and childcare are neglected.

In other words, we are increasingly producing and consuming much more value than our economic indicators measure. This suggests that we need a new way of measuring output and productivity, since we are not sufficiently taking into account the value that is being produced in the economy.

This can be seen as part and parcel of the "beyond GDP" debate, which argues that GDP is simply not a sufficient measure of societal progress. It will be particularly important to revise the traditional growth and productivity numbers, since most of these new productivity gains will be achieved in a way that makes our world more environmentally sustainable. Indeed, the examples cited above are emblematic of the new "sharing economy" where we make better use of existing products rather than merely producing more "stuff", which while good for the GDP statistics, is not necessarily so for the planet.

What happens when robots turn white-collar

And while discussions of productivity and measurement remain somewhat theoretical, nothing can be more concrete than the potential impact on what is arguably most fundamental to our sense of economic worth: gainful employment.

Throughout the ages, technology has replaced human effort, which while good for productivity growth (as mentioned above) and growth overall, is disruptive for those workers who lose their jobs. And this is no longer just about repetitive factory jobs: new computing and robotics technologies now threaten many professions that had seemed "safe territory", such as accountants, taxi drivers and paralegals.

Given the speed and breadth of the changes now being unleashed, it is clear that new technologies will dramatically change the nature of work across all industries and occupations. And as automation will inevitably replace labour in providing existing goods and services, the main question is how long this will take and how far it will go. A recent study estimated that 47% of total employment in the US is at risk, over the next decade or two.

It has always been the case that technological innovation destroys some jobs and replaces them in turn with new ones, in a different activity and possibly in a different place. As technological innovation forges ahead, one can expect that low-skill activities will be progressively replaced by tasks that require creativity and social intelligence. And as the job market becomes increasingly segregated into "low-skill/pay" and "high-skill/pay" segments, social tensions will inevitably rise.

We have already seen an increase of inequality within most OECD countries in recent decades, and institutions such as the IMF and the OECD are quantifying the extent to which this inequality is hampering growth and development.

School-work-retirement – RIP

Given that the dislocation will be significant and that the transition between the old and the new jobs will take time, the main question is what to do to foster more positive outcomes and best manage those caught in the transition. In a working environment that evolves so rapidly, the ability to anticipate future requirements in terms of the knowledge and the skills necessary to adapt becomes increasingly critical.

All stakeholders – business, government, society, and individuals – will have to work together to adjust education and training systems that can continuously reskill and "up-skill" workers. The traditional model of school-work-retirement will simply not cut it any more. This will be particularly important if we are entering an era when jobs are being rendered obsolete much faster than new ones are created.

Will developing countries leapfrog ahead – or be left behind?

Finally, it is important to reflect upon what this might mean for developing countries. Given that many of the past phases of the industrial revolution have not yet reached many of the world's citizens (who still do not have access to electricity, tractors, etc.), the Fourth Industrial Revolution mainly characterizes what is transpiring in the advanced (and to a certain extent middle-income) economies.

Over recent decades, although there has been a rise in inequality within countries, inequality across countries decreased significantly as developing countries began to catch up. Does it risk potentially reversing the catch up we have seen to date in terms of income, skills, infrastructure, finance, etc.? Or on the other hand, will these technologies and rapid changes be harnessed for development and faster catch up through leapfrogging?

The *homo economicus* of tomorrow

It is hard to answer these questions, but they will require significant thought as advanced economies contend with their own challenges. It is not only a moral imperative to ensure that swathes of the globe are not left behind; such a scenario would also pose a risk to global stability through channels such as global inequality, migration flows, and even geopolitical relations and security.

Ultimately, developing countries have the greatest gap to close, but can also benefit from learning from the mistakes of the advanced economies, leapfrogging to more prosperous and technologically enhanced futures. The successful homo economicus of tomorrow will certainly be different from today: she will be highly creative and adaptive; she will have many jobs in her lifetime where unthought-of technologies support her extreme efficiency, and she will probably not bother to own (or drive) a car to work, in the case that she even has a physical office to go to. She will live in a world that has been profoundly altered by the Fourth Industrial Revolution. Now is the time to make sure it is changed for the better.

Robots in War: The Next Weapons of Mass Destruction

Imagine turning on the TV and seeing footage from a distant urban battlefield, where robots are hunting down and killing unarmed children. It might sound like science fiction – a scene from *Terminator*, perhaps. But unless we act soon, lethal autonomous weapons systems – robotic weapons that can locate, select, and attack human targets without human intervention – could be a reality.

I'm not the only person worried about a future of "killer robots". In July, over 3,000 of researchers in artificial intelligence and robotics research – including many members of the Forum's Global Agenda Council on Artificial Intelligence and Robotics – signed an open letter calling for a treaty to ban lethal autonomous weapons. We were joined by another 17,000 signatories from fields as diverse as physics, philosophy, and law, including Stephen Hawking, Elon Musk, Steve Wozniak and Noam Chomsky.

The letter made headlines around the world, with more than 2,000 media articles in over 50 countries. So, what is the debate about?

A new breed of robots

We're all familiar, in varying degrees, with three pieces of modern technology:

1. The self-driving car: You tell it where to go and it chooses a route and does all the driving, "seeing" the road through its onboard camera.

2. Chess software: You tell it to win and it chooses where to move its pieces and which enemy pieces to capture.

3. The armed drone: You fly it remotely through a video link; you choose the target; and you launch the missile.

A lethal autonomous weapon might combine elements of all three: imagine that instead of a human controlling the armed drone, the chess software does, making its own tactical decisions, and using vision technology from the self-driving car to navigate and recognize targets.

In the UK – one of at least six states researching, developing and testing fully autonomous weapons – the Ministry of Defense has said that such weapons are probably feasible now for some aerial and naval scenarios. Two programs from the US Defense Advanced Research Project Agency (DARPA) already provide clues as to how autonomous weapons might be used in urban settings. The Fast Lightweight Autonomy program will see tiny rotorcraft maneuver unaided at high speed in urban areas and inside buildings. The Collaborative Operations in Denied Environment program plans to create teams of autonomous aerial vehicles that could carry out every step of a strike mission in situations where enemy jamming makes communication with a human commander impossible. In a press release describing the program DARPA memorably compared it to "wolves hunting in coordinated packs".

There is no doubt that as the technology improves, autonomous weapons will be highly effective. But does that necessarily mean they're a good idea?

Robotic outlaws

We might think of war as a complete breakdown of the rule of law, but it does in fact have its own legally recognized codes of conduct. Many experts in this field, including Human Rights Watch, the International

Committee of the Red Cross and UN Special Rapporteur Christof Heyns, have questioned whether autonomous weapons could comply with these laws. Compliance requires subjective and situational judgements that are considerably more difficult than the relatively simple tasks of locating and killing – and which, with the current state of artificial intelligence, would probably be beyond a robot.

Even those countries developing fully autonomous weapons recognize these limitations. In 2012, for example, the US Department of Defense issued a directive stating that such weapons must be designed in a way that allows operators to "exercise appropriate levels of human judgement over the use of force". The directive specifically prohibits the autonomous selection of human targets, even in defensive settings.

Life-saving robot warriors

But some robotics experts think that lethal autonomous weapons systems could actually reduce the number of civilian wartime casualties. The argument is based on an implicit *ceteribus paribus* assumption that, after the advent of autonomous weapons, the specific killing opportunities – numbers, times, locations, places, circumstances, victims – will be exactly the same as would have occurred with human soldiers.

This is rather like assuming cruise missiles will only be used in exactly those settings where spears would have been used in the past. Obviously, autonomous weapons are completely different from human soldiers and would be used in completely different ways – for example, as weapons of mass destruction.

Moreover, it seems unlikely that military robots will always have their "humanitarian setting" at 100%. One cannot consistently claim that the well-trained soldiers of civilized nations are so *bad* at following the rules of war that robots can do better, while at the same time claiming that rogue nations, dictators, and terrorist groups are so *good* at following the rules of war that they will never use robots in ways that violate these rules.

Beyond these technological issues of compliance, there are fundamental moral questions. The Martens Clause of the Geneva Conventions – a set of treaties that provide the framework for the law of armed conflict – declares that, "The human person remains under the protection of the principles of humanity and the dictates of public conscience." It's a sentiment echoed by countries such as Germany, which has said it "will not accept that the decision over life and death is taken solely by an autonomous system".

At present, the public has little understanding of the state of technology and the near-term possibilities, but this will change once video footage of robots killing unarmed civilians starts to emerge. At that point, the dictates of public conscience will be very clear but it may be too late to follow them.

Robots of mass destruction

The primary strategic impact of autonomous weapons lies not so much in combat superiority compared to manned systems and human soldiers, but in their *scalability*. A system is scalable if one can increase its impact just by having lots more of it; for example, as we scale nuclear bombs from tons to kilotons to megatons, they have much more impact. We call them weapons of mass destruction for a good reason. Kalashnikovs are not scalable in the same sense. A million Kalashnikovs can kill an awful lot of people, but only if carried by a million soldiers, who require a huge military-industrial complex to support them – essentially a whole nation-state.

A million autonomous weapons, on the other hand, need just a few people to acquire and program them – no human pilots, no support personnel, no medical corps. Such devices will form a new, scalable class of weapons of mass destruction with destabilizing properties like those of biological weapons: they tip the balance of power away from legitimate states and towards terrorists, criminal organizations, and other non-state actors. Finally, they are well suited for repression, being immune to bribery or pleas for mercy.

Autonomous weapons, unlike conventional weapons, could also lead to strategic instability. Autonomous weapons in conflict with other autonomous weapons *must* adapt their behaviour quickly, or else their predictability leads to defeat. This adaptability is necessary but makes autonomous weapons intrinsically unpredictable and hence difficult to control. Moreover, the strategic balance between robot-armed countries can change overnight thanks to software updates or cybersecurity penetration. Finally, many military analysts worry about the possibility of an accidental war – a military "flash crash".

Where now?

The UN has already held several meetings in Geneva to discuss the possibility of a treaty governing autonomous weapons. Ironing out the details of a treaty will be quite a challenge, though not impossible. Perhaps more complicated are the issues of treaty verification and diversion of dual-use technology. Experience with the Biological and Chemical Weapons Conventions suggests that transparency and industry cooperation will be crucial.

The pace of technological advances in the area of autonomy seems to be somewhat faster than the typical process of creating arms-control treaties – some of which have been many decades in development. The process at present is balanced on a knife-edge: while many nations have expressed strong reservations about autonomous weapons, others are pressing ahead with research and development. International discussions over the next couple of years will be crucial. Time is of the essence.

Ensuring AI Benefits Humanity?

The robots are coming, and Silicon Valley wants to make sure they're on our side. Tesla and SpaceX CEO Elon Musk, along with a veritable cornucopia of technology giants and scientists, announced the formation of OpenAI, a "non-profit artificial intelligence research company."

According to the company's press release, Musk will lead the non-profit, along with Sam Altman, the founder of startup incubator Y Combinator. The two are joined by backing from Reid Hoffman, the founder of LinkedIn; Jessica Livingston, another Y Combinator founder; and Peter Thiel, who cofounded PayPal with Musk and is now a prominent venture capitalist. The group, along with Amazon Web Services, Infosys, and Y Combinator, have pledged $1 billion for artificial intelligence research that keeps humanity at its core.

Musk, along with many other prominent tech figures, has in the past voiced his concerns over the way artificial intelligence is being researched. He signed a letter aiming to ensure we don't develop AI weapons, and he previously committed $10 million to the Future of Life Institute, another non-profit aiming to ensure that AI doesn't negatively affect humanity. OpenAI appears to be an extension of his long-held feelings on the future of artificial intelligence.

Altman told Quartz that he and others in the group had been brainstorming the idea for a while. "Over the last couple of months, it really started to come together," he said. "We were able to recruit what I think its one of the best collections of researchers in the world: We'll get started and see how it goes."

Many have suggested that once we develop a truly artificially intelligent system, it could quickly blow by our own intelligence. "It's hard to fathom how much human-level AI could benefit society, and it's equally hard to imagine how much it could damage society if built or used incorrectly," the group said in its release.

Some prominent AI researchers, including Andrej Karpathy, and Wojciech Zaremba and Pamela Vagata, both of whom work with Facebook's lead AI researcher, Yann LeCun, join the group of engineers and technologists. OpenAI said in its release that it wants to grow into a research institution that will share with the wider AI research community. While OpenAI won't be able to directly influence the work of the entire AI community, Altman said that because the group has "no fiduciary duty to shareholders," he hopes it will be able to collaborate with a wide swath of industry and academic groups researching in the field.

Many companies are pouring funds into AI research, from Silicon Valley firms like Google, to industrial titans like Toyota, which also recently pledged $1 billion to AI and robotics research. "I'd say hopefully we'll do a better job spending that than Toyota does," Altman said.

Mind Control: Man-Machine Gap Blurred

Imagine a condition that leaves you fully conscious, but unable to move or communicate, as some victims of severe strokes or other neurological damage experience. This is locked-in syndrome, when the outward connections from the brain to the rest of the world are severed. Technology is beginning to promise ways of remaking these connections, but is it our ingenuity or the brain's that is making it happen?

Ever since an 18th-century biologist called Luigi Galvani made a dead frog twitch we have known that there is a connection between electricity and the operation of the nervous system. We now know that the signals in neurons in the brain are propagated as pulses of electrical potential, whose effects can be detected by electrodes in close proximity. So, in principle, we should be able to build an outward neural interface system – that is to say, a device that turns thought into action.

In fact, we already have the first outward neural interface system to be tested in humans. It is called BrainGate and consists of an array of micro-electrodes, implanted into the part of the brain concerned with controlling arm movements. Signals from the micro-electrodes are decoded and used to control the movement of a cursor on a screen, or the motion of a robotic arm.

A crucial feature of these systems is the need for some kind of feedback. A patient must be able to see the effect of their willed patterns of thought on the movement of the cursor. What's remarkable is the ability of the brain to adapt to these artificial systems, learning to control them.

Virtual reality

Inward neural interfaces – ones that provide inputs to the brain – also depend on the brain's ability to adapt to them. Cochlear implants, which can restore some measure of hearing to the profoundly deaf, have been around for several decades now. These take signals from an external microphone, and after signal processing, transmit a series of pulses to electrodes that excite the auditory nerve. The pulses are designed to mimic the way different frequencies are encoded by a functioning cochlea, but the match is imperfect, and the restoration of the ability to understand speech, for example, depends on the brain's impressive ability to learn to adapt to the new kinds of input.

The first trials of retinal implants have now taken place, in which signals from a camera are used to stimulate retinal neurons in vision-impaired patients. Second Sight's Argus II system shows some encouraging results, with patients able to pick out shapes and detect the motion of objects. For the first time, people who have become blind due to the degeneration of their own photoreceptor cells – which convert light into signals in the eyes – can have some measure of artificial vision restored.

The key message of all this is that brain interfaces now are a reality and that the current versions will undoubtedly be improved. In the near future, for many deaf and blind people, for people with severe disabilities – including, perhaps, locked-in syndrome – there are very real prospects that some of their lost capabilities might be at least partially restored.

Until then, our current neural interface systems are very crude. One problem is size; the micro-electrodes in use now, with diameters of tens of microns, may seem tiny, but they are still coarse compared to the sub-micron dimensions of individual nerve fibers. And there is a problem of scale. The BrainGate system, for example, consists of 100 micro-electrodes in a square array; compare that to the many tens of billions of neurons in the brain. The fact these devices work at all is perhaps more a testament to the adaptability of the human brain than to our technological prowess.

Scale models

So, the challenge is to build neural interfaces on scales that better match the structures of biology. Here, we move into the world of nanotechnology. There has been much work in the laboratory to make nano-electronic structures small enough to read out the activity of a single neuron. In the 1990s, Peter Fromherz, at the Max Planck Institute for Biochemistry, was a pioneer of using silicon field effect transistors, similar to those used in commercial microprocessors, to interact with cultured neurons. In 2006, Charles Lieber's group at Harvard succeeded in using transistors made from single carbon nanotubes – whiskers of carbon just one nanometer in diameter – to measure the propagation of single nerve pulses along the nerve fibers.

But these successes have been achieved, not in whole organisms, but in cultured nerve cells which are typically on something like the surface of a silicon wafer. It's going to be a challenge to extend these methods into three dimensions, to interface with a living brain. Perhaps the most promising direction will be to create a 3D "scaffold" incorporating nano-electronics, and then to persuade growing nerve cells to infiltrate it to create what would in effect be cyborg tissue – living cells and inorganic electronics intimately mixed.

This prospect might be achievable in our lifetimes, but what does remain very far away is the transhumanist dream of being able to obtain a complete readout of the brain – a transcript of the state of the mind. Neural interfaces will remain only the narrowest and most partial of windows on the huge complexity of the inner life of a brain, though even that partial window will be life-transforming for some.

As brain interfaces improve, they will bring real benefits to many, and some ethical issues too. As the techniques become more routine, it's likely that people will find non-medical uses for them. We might find ourselves controlling computer games, or taking direct control of machines at work. We will still be a long way from the seamless integration of humans and machines, but the science fiction vision of the cyborg will become real enough to give us pause for thought.

Ethical Issues of Living with Robots

Even if robots are just tools, people will see them as more than that. For a sci-fi fan like me, fascinated by the nature of human intelligence and the possibility of building life-like robots, it's always interesting to identify a new angle on these questions. As a re-imagining of the original 1970s science fiction film set in a cowboy-themed, hyper-real adult theme park populated by robots that look and act like people, Westworld does not disappoint.

Westworld challenges us to consider the difference between being human and being a robot. From the beginning of this new serialization on HBO we are confronted with scenes of graphic human-on-robot violence. But the robots in Westworld have more than just human-like physical bodies: they display emotion including extreme pain; they see and recognize each other's suffering; they bleed and even die. What makes this acceptable, at least within Westworld's narrative, is that they are just extremely life-like human simulations; while their behaviour is realistically automated, there is "nobody home".

But from the start, this notion that a machine of such complexity is still merely a machine is undermined by constant reminders that they are also so much like us. The disturbing message, echoing that of previous sci-fi classics such as Blade Runner and AI, is that machines could one day be so close to human as to be indistinguishable – not just in intellect and appearance, but also in moral terms.

At the same time, by presenting an alternate view of the human condition through the technological mirror of life-like robots, Westworld causes us to reflect that we are perhaps also just sophisticated machines, albeit of a biological kind – an idea that has been forcefully argued by the philosopher Daniel Dennett.

The unfortunate robots in Westworld have, at least initially, no insight into their existential plight. They enter each new day programmed with enthusiasm and hope, oblivious to its pre-scripted violence and tragedy. We may pity these automatons their fate – but how closely does this blinkered ignorance, belief in convenient fictions, and misguided presumption of agency resemble our own human predicament?

Westworld arrives at a time when people are already worried about the real-world impact of advances in robotics and artificial intelligence. Physicist Stephen Hawking and technologist Elon Musk are among the powerful and respected voices to have expressed concern about allowing the AI genie to escape the bottle. Westworld's contribution to the expanding canon of science fiction dystopias will do nothing to quell such fears. Channeling Shakespeare's King Lear, a malfunctioning robot warns us in chilling terms: "I shall have such revenges on you both. The things I will do, what they are, yet I know not. But they will be the terrors of the Earth."

But against these voices are other distinguished experts trying to quell the panic. For Noam Chomsky, the intellectual godfather of modern AI, all talk of matching human intelligence in the foreseeable future remains fiction, not science. One of the world's best-known roboticists, Rodney Brooks has called on us to relax: AI is just a tool, not a threat.

We are far from being able to replicate human intelligence in robot form. Our current systems are too simple, probably by several orders of magnitude. Building human-level AI is extremely hard; as Brooks says, we are just at the beginning of a very long road. But I see the path along which we are developing AI as one of symbiosis, in which we can use robots to benefit society and exploit advances in artificial intelligence to boost our own biological intelligence.

More than just a tool

Nevertheless, in recent years the robots and AI are "just tools" line of argument is frustrating: partly because it has failed to calm the disquiet around AI, and partly because there are good reasons why these technologies are different from others in the past.

Even if robots are just tools, people will see them as more than that. It seems natural for people to respond to robots – even some of the more simple, non-human robots we have today – as though they have goals and intentions. It may be an innate tendency of our profoundly social human minds to see entities that act intelligently in this way. More importantly, people may see robots as having psychological properties such as the ability to experience suffering.

It may be difficult to persuade them to see otherwise, particularly if we continue to make robots more life-like. If so, we may have to adapt our ethical frameworks to take this into account. For instance, we might consider violence towards a robot as wrong, even though the suffering is imagined rather than real. Indeed, faced with violence towards a robot some people show this sort of ethical response spontaneously. We will have to deal with these issues as we learn to live with robots.

As AI and robot technology becomes more complex, robots may come to have interesting psychological properties that make them more than just tools. The fictional robots of Westworld are clearly in this category. But already real robots are being developed that have artificial drives and motivations; that are aware of their own bodies as distinct from the rest of the world; that are equipped with internal models of themselves and others as social entities, and that are able to think about their own past and future.

These are not properties that we find in drills and screwdrivers. They are embryonic psychological capacities that, so far, have only been found in living, sentient entities such as humans and animals. Stories like that of Westworld remind us that as we progress toward ever more sophisticated AI, it might lead us to machines that are like us, and we also see ourselves as machines.

Ethical Issues in Artificial Intelligence

Faced with an automated future, what moral framework should guide us? Optimizing logistics, detecting fraud, composing art, conducting research, providing translations: intelligent machine systems are transforming our lives for the better. As these systems become more capable, our world becomes more efficient and consequently richer.

Tech giants such as Alphabet, Amazon, Facebook, IBM, and Microsoft – as well as individuals like Stephen Hawking and Elon Musk – believe that now is the right time to talk about the nearly boundless landscape of artificial intelligence. In many ways, this is just as much a new frontier for ethics and risk assessment as it is for emerging technology. So which issues and conversations keep AI experts up at night?

1. Unemployment. What happens after the end of jobs?

The hierarchy of labour is concerned primarily with automation. As we've invented ways to automate jobs, we could create room for people to assume more complex roles, moving from the physical work that dominated the pre-industrial globe to the cognitive labour that characterizes strategic and administrative work in our globalized society.

Look at trucking: it currently employs millions of individuals in the United States alone. What will happen to them if the self-driving trucks promised by Tesla's Elon Musk become widely available in the next decade? But on the other hand, if we consider the lower risk of accidents, self-driving trucks seem like an ethical choice. The same scenario could happen to office workers, as well as to the majority of the workforce in developed countries.

This is where we come to the question of how we are going to spend our time. Most people still rely on selling their time to have enough income to sustain themselves and their families. We can only hope that this opportunity will enable people to find meaning in non-labour activities, such as caring for their families, engaging with their communities and learning new ways to contribute to human society.

If we succeed with the transition, one day we might look back and think that it was barbaric that human beings were required to sell the majority of their waking time just to be able to live.

2. Inequality. How do we distribute the wealth created by machines?

Our economic system is based on compensation for contribution to the economy, often assessed using an hourly wage. The majority of companies are still dependent on hourly work when it comes to products and services. But by using artificial intelligence, a company can drastically cut down on relying on the human workforce, and this means that revenues will go to fewer people. Consequently, individuals who have ownership in AI-driven companies will make all the money.

We are already seeing a widening wealth gap, where start-up founders take home a large portion of the economic surplus they create. In 2014, roughly the same revenues were generated by the three biggest companies in Detroit and the three biggest companies in Silicon Valley ... only in Silicon Valley there were 10 times fewer employees.

If we're truly imagining a post-work society, how do we structure a fair post-labour economy?

3. Humanity. How do machines affect our behaviour and interaction?

Artificially intelligent bots are becoming better and better at modelling human conversation and relationships. In 2015, a bot named Eugene Goostman won the Turing Challenge for the first time. In this challenge, human raters used text input to chat with an unknown entity, then guessed whether they had been chatting with a human or a machine. Eugene Goostman fooled more than half of the human raters into thinking they had been talking to a human being.

This milestone is only the start of an age where we will frequently interact with machines as if they are humans; whether in customer service or sales. While humans are limited in the attention and kindness that they can expend on another person, artificial bots can channel virtually unlimited resources into building relationships.

Even though not many of us are aware of this, we are already witnesses to how machines can trigger the reward centres in the human brain. Just look at click-bait headlines and video games. These headlines are often optimized with A/B testing, a rudimentary form of algorithmic optimization for content to capture our attention. This and other methods are used to make numerous video and mobile games become addictive. Tech addiction is the new frontier of human dependency.

On the other hand, maybe we can think of a different use for software, which has already become effective at directing human attention and triggering certain actions. When used right, this could evolve into an opportunity to nudge society towards more beneficial behavior. However, in the wrong hands it could prove detrimental.

4. Artificial stupidity. How can we guard against mistakes?

Intelligence comes from learning, whether you're human or machine. Systems usually have a training phase in which they "learn" to detect the right patterns and act according to their input. Once a system is fully trained, it can then go into test phase, where it is hit with more examples and we see how it performs.

Obviously, the training phase cannot cover all possible examples that a system may deal with in the real world. These systems can be fooled in ways that humans wouldn't be. For example, random dot patterns can lead a machine to "see" things that aren't there. If we rely on AI to bring us into a new world of labour, security and efficiency, we need to ensure that the machine performs as planned, and that people can't overpower it to use it for their own ends.

5. Racist robots. How do we eliminate AI bias?

Though artificial intelligence is capable of a speed and capacity of processing that's far beyond that of humans, it cannot always be trusted to be fair and neutral. Google and its parent company Alphabet are one of the leaders when it comes to artificial intelligence, as seen in Google's Photos service, where AI is used to identify people, objects and scenes. But it can go wrong, such as when a camera missed the mark on racial sensitivity, or when a software used to predict future criminals showed bias against black people.

We shouldn't forget that AI systems are created by humans, who can be biased and judgmental. Once again, if used right, or if used by those who strive for social progress, artificial intelligence can become a catalyst for positive change.

6. Security. How do we keep AI safe from adversaries?

The more powerful a technology becomes, the more can it be used for nefarious reasons as well as good. This applies not only to robots produced to replace human soldiers, or autonomous weapons, but to AI systems that can cause damage if used maliciously. Because these fights won't be fought on the battleground only, cybersecurity will become even more important. After all, we're dealing with a system that is faster and more capable than us by orders of magnitude.

7. Evil genies. How do we protect against unintended consequences?

It's not just adversaries we have to worry about. What if artificial intelligence itself turned against us? This doesn't mean by turning "evil" in the way a human might, or the way AI disasters are depicted in Hollywood movies. Rather, we can imagine an advanced AI system as a "genie in a bottle" that can fulfill wishes, but with terrible unforeseen consequences.

In the case of a machine, there is unlikely to be malice at play, only a lack of understanding of the full context in which the wish was made. Imagine an AI system that is asked to eradicate cancer in the world. After a lot of computing, it spits out a formula that does, in fact, bring about the end of cancer — by killing everyone on the planet. The computer would have achieved its goal of "no more cancer" very efficiently, but not in the way humans intended it.

8. Singularity. How do we stay in control of a complex intelligent system?

The reason humans are on top of the food chain is not down to sharp teeth or strong muscles. Human dominance is almost entirely due to our ingenuity and intelligence. We can get the better of bigger, faster, stronger animals because we can create and use tools to control them: both physical tools such as cages and weapons, and cognitive tools like training and conditioning.

This poses a serious question about artificial intelligence: will it, one day, have the same advantage over us? We can't rely on just "pulling the plug" either, because a sufficiently advanced machine may anticipate this move and defend itself. This is what some call the "singularity": the point in time when human beings are no longer the most intelligent beings on earth.

9. Robot rights. How do we define the humane treatment of AI?

While neuroscientists are still working on unlocking the secrets of conscious experience, we understand more about the basic mechanisms of reward and aversion. We share these mechanisms with even simple animals. In a way, we are building similar mechanisms of reward and aversion in systems of artificial intelligence. For example, reinforcement learning is similar to training a dog: improved performance is reinforced with a virtual reward.

Right now, these systems are fairly superficial, but they are becoming more complex and life-like. Could we consider a system to be suffering when its reward functions give it negative input? What's more, so-called genetic algorithms work by creating many instances of a system at once, of which only the most successful "survive" and combine to form the next generation of instances. This happens over many generations and is a way of improving a system. The unsuccessful instances are deleted. At what point might we consider genetic algorithms a form of mass murder?

Once we consider machines as entities that can perceive, feel and act, it's not a huge leap to ponder their legal status. Should they be treated like animals of comparable intelligence? Will we consider the suffering of "feeling" machines?

Some ethical questions are about mitigating suffering, some about risking negative outcomes. While we consider these risks, we should also keep in mind that, on the whole, this technological progress means better lives for everyone. Artificial intelligence has vast potential, and its responsible implementation is up to us.

Japan's Robot Babies: are They Ethical?

Driven by a declining population, a trend for developing robotic babies has emerged in Japan as a means of encouraging couples to become "parents". The approaches taken vary widely and are driven by different philosophical approaches that also beg a number of questions, not the least whether these robo-tots will achieve the aim of their creators.

To understand all of this it is worth exploring the reasons behind the need to promote population growth in Japan. The issue stems from the disproportionate number of older people. Predictions from the UN suggest that by 2050 there will be about double the number of people living in Japan in the 70-plus age range compared to those aged 15-30. This is blamed on a number of factors including so-called "parasite singles", more unmarried women and a lack of immigration.

So, what are the different design approaches that are being taken to encourage more people to become parents? These have ranged from robots that mimic or represent the behaviour of a baby through to robots that look much more lifelike. Engineers at Toyota recently launched Kirobo Mini, for example, as a means of promoting an emotional response in humans. The robot does not look like a baby, but instead models "vulnerable" baby-like behaviours including recognising and responding to people in a high-pitched tone and being unstable in its movements.

At the other end of the spectrum is Yotaro, a robotic baby simulator that uses projection technology for its face so it can simulate emotions and expressions. The simulator also models reaction to touching, mood and even illness through an in-built runny nose.

Encouraging or off-putting?

Past evidence might suggest that giving couples robotic baby simulators would encourage population growth. Recent educational experiments with robotic babies and teenagers in the US and Australia, for example, found that although robotic babies were tested as a means of deterring teenage pregnancies actually increased among those groups that were allocated robotic babies compared to control groups.

However, it would be too simplistic to say that this might be the same result for all adopters of robotic babies. Ages and cultural differences would play a significant part in any outcome.

As well as aiming to promote a growth in population, researchers are also aiming to prepare young couples for the longer term needs of a child as it grows. Robots have been developed to represent children in a range of age groups, from "nine-month-old" Noby to "two-year-old" toddlers such as CB2 (although the latter is the output of research exploring the development of a biometric body).

While much focus has been on what goes into a baby robot, there are potential emotional issues for "parents". There have been a number of studies that have examined the relationship between humans and robots. Researchers have discovered a high degree of bonding can form between the two, a bond that is strengthened when the device is a social robot which may have a human-like appearance or portray human-like behaviours.

There are some interesting caveats to this rule of thumb, such as the "uncanny valley" identified by Mashiro Mori, which suggests there is a range of realistic human qualities that humans find repulsive rather than appealing.

At present, development is very much a one-way relationship; one in which the human projects human qualities onto the robot. But there are currently a number of projects underway to develop robots that make use of Artificial Intelligence techniques so that they can form their own relationships with humans.

This then leads to the ethical implications of using robots. Embracing a number of areas of research, robot ethics considers whether the use of a device within a particular field is acceptable and also whether the device itself is behaving ethically. When it comes to robot babies there are already a number of issues that are apparent. Should "parents" be allowed to choose the features of their robot, for example? How might parents be counselled when returning their robot baby? And will that baby be used again in the same form?

These problems may persist throughout the lifespan of the "child". If a point in time arrives when parents need to swap their robot baby for another due to defects or because they want an older "child", for example, how might the emotional attachment to the first "child" migrate to the replacement given that this really should be the evolution of the same "person"? In practical terms, this may be possible through software updates similar to updates to apps on smartphones today – or even transplanting components to allow the evolved "child" to retain characteristics and memories, similar to replacing a hard disk drive in a computer.

Even taking Asimov's "three laws" of robotics into account becomes problematic depending upon the interpretation of the laws. For example, the first law states that a robot should not harm a human being. What if harm can be considered as emotional or psychological? You could argue that a human may be emotionally harmed when bonding with a robot baby as a result of the robot's actions.

The use of social robots in general raises many issues, both ethical and technical. The problem of declining birthrates is, however, a real and growing problem in a number of nations. Robot babies may not directly prove to be a solution, but it may lead to research that offers better understanding and insight into the problem of birthrate decline.

Can Algorithm Choose the Next POTUS?

Imagine a typical day in 2020: your personal AI assistant wakes you up with a friendly greeting before preparing your favourite breakfast. During your morning workout, they play new songs that perfectly match your taste. For your commute to work, they've already pre-selected a few articles based on the duration of your commute and what you've read in the past.

You read the news and remember the presidential elections are coming up. Based on a predicted model that considers your previously expressed views and data on other voters in your state, your AI assistant recommends you vote Democrat. A pop-up message on your phone asks whether you want your AI assistant to handle the paperwork and cast the vote on your behalf. You tap "agree" and get on with your life.

AI: only as good as the data

While personal AI assistants are already becoming a reality, to many it seems inconceivable that we would ever delegate such important civic duties – even if an AI assistant probably knows what's best for us at any given moment. If fed enough data, an AI assistant could give recommendations that are far more accurate and personalized than we'd receive from even our closest friends.

Alphabet's Chairman Eric Schmidt is convinced that advances in AI will make each and every human being in the world smarter, more capable, and more accomplished. There is hope that AI can help society solve some of its toughest challenges, including climate change, population growth and human development.

Yet the intelligent potential of machines frequently elicits fear in people. Studies show that 34% of people are afraid of AI, while 24% think AI will be harmful for society. GWI finds that 63% of people worry about how their personal data is used by companies. The recent research at the Oxford Internet Institute shows that people hesitate to put their personal lives in the hands of an AI assistant, especially when that assistant makes decisions without providing a transparent reasoning for choosing one solution over a set of alternatives.

There is no need to confuse math with magic. AIs are not operated by small mystical creatures living inside your phones that have a sense of agency and their own agenda. But what we tend to forget is that the seemingly invisible mathematical models that make automated inferences about our interests, locations, behaviours, finances and health are designed by other humans using our pre-existing personal data.

Role of human creators

Much of the current debate on algorithmic culture revolves around the role that humans play in the design of algorithms – that is, whether a creator's subconscious beliefs and biases are encoded into the algorithms that make decisions about us. Journalism is replete with fears that developers willingly code their algorithms in a way that would permit subtle discrimination against certain groups of people over others – or worse, that technology platforms act as gatekeepers for the information that passes through them. As John Mannes writes, "A biased world can result in biased datasets and, in turn, bias artificial intelligence frameworks."

The policy-makers and pundits advancing this argument tend to misconstrue the evidence for the apparent prevalence of algorithmic biases. They blame the people behind the algorithms, rather than the algorithms themselves. Finding fault in other people is, of course, the natural response, especially if you don't understand the inner workings of the technologies at hand.

But algorithmic bias rarely originates from its human creators. In the majority of cases, it originates from the data that is used to train such algorithms. And this is, indeed, the real danger of the futuristic scenario described at the outset of this article.

Algorithmic determinism

Let's pause for a moment and recall how machine learning actually works. By applying statistical learning techniques, we can develop computational models that can automatically identify patterns in the data. To accomplish that, the models need to be trained on large datasets to unearth boundaries and relationships in the data. The more data is used to train the model, the higher the predictive accuracy.

In the context of personalized digital applications, these statistical learning techniques are used to create an algorithmic identity for their users, which encompasses several dimensions, such as use patterns, tastes, preferences, personality traits and the structure of their social graph. This digital identity, however, is not directly based on users' personhood or sense of self. Rather, such inferences are based on a collection of measurable data points and the machine's interpretation thereof. In other words, the embodied user identity, no matter how complex, is replaced by an imperfect digital representation of itself in the eyes of the AI.

But the AI can only use historically traced data in its computations, which is then used to anticipate the users' needs and make predictions about the future. This is why neural networks trained on images of past US presidents predicted Donald Trump would win the upcoming US election, after being trained with images of past (male) presidents. Since there were no female US presidents in the dataset, the AI was unable to deduce that gender was not a relevant characteristic for the model. In practice, if this particular AI were to elect the next president, it would vote for Trump.

These inferences result in increasingly deterministic recommendation systems, which tend to reinforce existing beliefs and practices similar to the echo chambers in our social media feeds. At this point, Jarno Koponen asserts that "personalization caricaturizes us and creates a striking gap between our real interests and their digital reflection". Meanwhile, the authors of *The Netflix Effect* explain that personal recommendation systems tend to "steer the user towards this content, thus ghettoizing the user in a prescribed category of demographically classified content." These aspects become more prevalent over time and it becomes evident why algorithmic determinism may be so harmful.

Fluid identities and change

Our identities are dynamic, complex and contain many contradictions. Based on our social context, we may behave differently and require different forms of assistance from our AI agents – at school or at work, in a bar or in church.

Alongside our default self-presentation, there are many reasons why we may want to enact different identities to differentially engage with various sub-groups in our personal networks. Do we want to make ourselves socially accessible to our entire social network, or do we want to find new untapped social spaces away from the prying eyes of our friends and family? What happens if we want to **experiment with different social roles and facets of our identity? As 4Chan founder** Chris Poole, a.k.a. mootm says, "it's not who you share with; it's who you share as (…). Identity is prismatic; there are many lenses through which people view you".

Distinguishing these different layers of self-presentation and mapping them onto various social environments is an extremely challenging task for an AI, which has been trained to serve a unique user identity. There are days when even we don't know who we are. But our AI assistants will always have an answer for us: it's who we were yesterday. Behaviour change becomes increasingly difficult, and our use patterns and belief systems run the risk of being locked up in a self-enforcing cycle, similar to an algorithmic Groundhog Day.

The more we rely on personalized algorithms in everyday life, the more they will shape what we see, what we read, who we talk to, and how we live. By relentlessly focusing on the status quo, new recommendations on books to read, movies to watch and people to meet will give us more of the same things that have previously delighted us. This is what algorithmic determinism is all about.

When your past unequivocally dictates your future, personal development through spontaneity, open-mindedness and experimentation becomes more difficult. In this way, the notion of algorithmic determinism echoes what Winston Churchill once said about buildings: We shape our algorithms; thereafter, they shape us.

How to Stop (or Slow Down) the Future?

Today, real-world applications of AI are already embedded in almost every aspect of everyday life – and interest in the technology is growing.

In terms of technological progress, there is a lack of inter-operability standards for data exchange between applications, which prevents radical personalization. To be truly useful, machine learning systems require greater amounts of personal data – data that is currently siloed in proprietary databases of competing technology companies. Those who have the data hold the power. Some companies, most notably Apple and Viv, have started to democratize this power by experimenting with third-party service integration. Most recently, some of the largest technology companies announced a major partnership to collaborate on AI research that benefits the many, not the few. Going forward, this will be crucial to establishing public trust in AI.

In terms of social progress, there is an implicit aversion to the rapid growth of AI. People are afraid of losing control of their AI assistants. Trust is a direct function of our ability to control something. Putting their relationships and reputation on the line for a seemingly marginal improvement in productivity is not something most people are willing to do.

Of course, in its early stages, an AI assistant can behave in ways that its human creators might not expect. There are plenty of precedents of failed AI experiments that have diminished trust in narrow AI solutions and conversational bots. When Facebook, Microsoft, and Google all launched their bot platforms in 2016, users were disappointed with the limited usefulness, application and customization of the prematurely presented AI technologies.

The lingering fears about the ramifications of AI technology have also been fueled by the many dystopian sci-fi scenarios that have depicted sentient rogue AIs taking over the world. But the future we're heading towards will be neither Skynet nor Orwellian Big Brother: it is much more likely to look like the hedonic society portrayed in *A Brave New World*, where technologies maintain the status qua through a regime of universal happiness and pervasive self-indulgence.

Future-oriented regulation

Technologies keep marching ahead, but there's also hope. The 2016 annual survey of the Global Shapers Community revealed that young people see AI as the main technological trend. What is more, 21% of respondents say they support rights for humanoid robots, with this support disproportionally higher in South-East Asia, where young people appear to have a more favorable attitude towards the role of AI in everyday life.

In Europe, the European Union's new General Data Protection Regulation could restrict extreme forms of algorithmic determinism by giving users the opportunity to ask for an explanation of particular profiling-based algorithmic decisions. This law is expected to be implemented in all EU member states by May 2018. While this regulation restricts profiling, and underlines the pressing importance of human interpretability in algorithm design, it is uncertain whether it will result in any major changes to the prevailing algorithmic practices of large technology companies.

Thousands of algorithmic decisions are made about each of us every day – ranging from Netflix movie recommendations and Facebook friend suggestions to insurance risk assessments and credit scores. All things considered, are citizens themselves now responsible for keeping track of and scrutinizing the algorithmic decisions made about them, or is this something that needs to be encoded into the designs of the digital platforms they're using? Accountability is an important issue here, precisely because it is so difficult to measure and implement on a large scale.

Thus, the question we ought to ask ourselves before leaping headlong into the unknown is what do we want the relationship between humans and AI to look like? Rethinking these issues will hopefully allow us to design non-deterministic, transparent, and accountable algorithms that recognize the complex, evolving and multifaceted nature of our individuality.

Star Wars & Loyalty of Robots

Artificial intelligence could be encouraged, reprogrammed or hacked to defect.

The latest Star Wars movie, Rogue One introduces us to a new droid K-2SO that is the robotic lead of the story.

Without giving away too many spoilers, K-2SO is part of the Rebellion freedom fighter group that are tasked with stealing the plans to the first Death Star, the infamous moon-sized battle station from the original Star Wars movie.

The significance of K-2SO is his back-story. K-2SO is an autonomous military robot that used to fight for the Rebellion's enemy – the Imperial Empire. He was captured and reprogrammed by the rebels and is now a core member of Rogue One group.

K-2SO is not the first robot to swap sides in a movie. Remember the Terminator's initial mission was to kill Sarah Connor in the first movie, before being reprogrammed in later movies to protect her and her son John Connor.

This does then raise the question of whether in real life a programmed military machine could be encouraged, reprogrammed or hacked to defect.

Soldiers swapping sides

The idea of human soldiers swapping sides during wars and conflicts is nothing new. There are numerous examples of soldiers surrendering and then announcing that they have information and would like to help and sometimes fight for their captors.

It is the information about battle plans and tactics that these defecting soldiers have that could potentially change the course of a battle or a military campaign.

One of the most famous defectors was General Benedict Arnold. Arnold was a general of the American Army during the American War of Independence, but he defected to the British Army and became a brigadier general. He led British forces against the Americans and retired to London after the war.

Weapons technology

The industrial revolution and the rise of mechanical weapons such as tanks, aircraft and submarines in the early 20th century changed the nature of defecting.

It was the development of more and more advanced weapons that gave a nation its advantage over its military rivals.

Stealing an enemy's new weapons was almost impossible and so it was up to defectors to deliver the plans of the new weapons or sometimes, examples of the actual weapons to the other side.

Martin Monti, of the United States Army Air Corps, defected to Italy during 1944 and handed over a photographic reconnaissance version of the P-38 Lightning aircraft to the Nazi military. He then joined the Nazi SS.

In 1976, during the Cold War, Viktor Belenko, flew his highly-secret MiG-25 jet fighter from the USSR to Japan.

NATO had long wanted to get the technical details of this aircraft as it was rumoured to be able to fly three times faster than the speed of sound.

Japan gave the US access to the MiG and Belenko was eventually granted citizenship of the US. The plane was stripped and analysed by the Americans who also had a copy of the aircraft's technical manual that Belenko had also brought with him.

Defectors not necessary

In the 21st century we have seen the development of remotely controlled systems for reconnaissance, surveillance and the delivery of weapons to targets. Such systems are likely to be very important in the future of defence capabilities.

As this equipment, does not require a person on-board, it means that human defectors or spies are no longer required to deliver this robotic hardware to the opposition.

It is impossible to know for sure when the first unmanned system was successfully captured. But because these systems rely on external radio commands and infrastructure, such as GPS, it is plausible that they can be taken over and captured and it has almost certainly already happened.

In 2011, a US Air Force drone came down in Iran and was recovered by the Iranian state. That aircraft was a highly secretive RQ-170 stealth drone and the Iranians claimed that they had "spoofed" the drone into landing in Iran by creating fake GPS signals.

Experts in the US doubted those claims, but however the drone was captured, Iran ended up with a nearly intact state-of-the-art stealth drone.
They put it on display to international media and stated that they would reverse engineer it and create their own version of this high-tech robotic surveillance aircraft. Iran

now appears to have a squadron of these stealth drones, all based on the original captured aircraft.

Trusting autonomous robots

An obvious way to prevent the claimed GPS-spoofing or other similar hacks is to create systems that are truly autonomous and do not require or use external communication systems.

Such robots should be immune to hacking once deployed on their missions. But the development and use of truly autonomous robot weapon systems is a controversial topic.

Competing in AI Age

Until recently, artificial intelligence (AI) was similar to nuclear fusion in unfulfilled promise. It had been around a long time but had not reached the spectacular heights foreseen in its infancy. Now, however, AI is realizing its potential in achieving human-like capabilities, so it is time to ask: How can business leaders harness AI to take advantage of the specific strengths of man and machine?

AI is swiftly becoming the foundational technology in areas as diverse as self-driving cars and financial trading. Self-learning algorithms are now routinely embedded in mobile and online services. Researchers have leveraged massive gains in processing power and the data streaming from digital devices and connected sensors to improve AI performance. And machines have essentially cracked speech and vision specifically and human communication generally. The implications are profound:

• Because they know how to speak, read text, and absorb and retain encyclopedic knowledge, machines can interact with people intuitively and naturally on a wide range of topics at considerable depth.

• Because they can identify objects and recognize optical patterns, machines can leave the virtual and join the real world.

A field that once disappointed its proponents is now striking remarkably close to home as it expands into activities commonly performed by humans. AI programs, for example, have diagnosed specific cancers more accurately than radiologists. No wonder that traditional companies in finance, retail, health care, and other industries have started to pour billions of dollars into the field.

Because AI systems think and interact, they are invariably compared to people. But while humans are fast at parallel processing (pattern recognition) and slow at sequential processing (logical reasoning), computers have mastered the former in narrow fields and are superfast in the latter. Just as submarines don't swim, machines solve problems and accomplish tasks in their own way.

Without further quantum leaps in processing power, machines will not reach artificial general intelligence (AGI): the combination of vastly different types of problem-solving capabilities—the hallmark of human intelligence. Today's robo-car, for example, doesn't exhibit what we would consider common sense, such as abandoning an excursion to assist a child who has fallen off her bicycle. But when properly applied, AI excels at performing many business tasks quickly, intelligently, and thoroughly.

Artificial intelligence is no longer an elective. It is critical for companies to figure out how humans and computers can play off each other's strengths as intertwined actors to create competitive advantage.

The Evolution of Competitive Advantage

In simpler times, a technology tool, such as Walmart's logistics tracking system in the 1980s, could serve as a source of advantage. AI is different. The naked algorithms themselves are unlikely to provide an edge. Many of them are in the public domain, and businesses can access open-source software platforms, such as Google's TensorFlow. OpenAI, a nonprofit organization started by Elon Musk and others, is making AI tools and research widely available. And many prominent AI researchers have insisted on retaining the right to publish their results when joining companies such as Baidu, Facebook, and Google.

Rather than scrap traditional sources of competitive advantage, such as position and capability, AI reframes them. Companies, then, need a fluid and dynamic view of their strengths. Positional advantage, for example, generally focuses on relatively static aspects that allow a company to win market share: proprietary assets, distribution networks, access to customers, and scale. These articles of faith have to be reimagined in the AI world.

Let's look at three examples of how AI shifts traditional notions of competitive advantage.

- **Data.**

AI's strongest applications are data-hungry. Pioneers in the field, such as Facebook, Google, and Uber, have each secured a "privileged zone" by gaining access to current and future data, the raw material of AI, from their users and others in ways that go far beyond traditional data harvesting. Their scale gives them the ability to run more training data through their algorithms and thus improve performance. In the race to leverage fully functional self-driving cars, for example, Uber has the advantage of collecting 100 million miles of fleet data daily from its drivers. This data will eventually inform the company's mobility services. Facebook and Google take advantage of their scale and depth to hone their ad targeting.

Not all companies can realistically aspire to be Facebook, Google, or Uber. But they do not need to. By building, accessing, and leveraging shared, rented, or complementary data sets, even if that means collaborating with competitors, companies can complement their proprietary assets to create their own privileged zone. Sharing is not a dirty word. The key is to build an unassailable and advantaged collection of open and closed data sources.

- **Customer Access.**

AI also changes the parameters of customer access. Well-placed physical stores and high-traffic online outlets give way to customer insights generated through AI. Major retailers, for example, can run loyalty, point-of-sale, weather, and location data through their AI engines to create personalized marketing and promotion offers. They can predict your route and appetite—before you are aware of them—and conveniently provide familiar, complementary, or entirely new purchasing options. The suggestive power of many of these offers has generated fresh revenue at negligible marginal cost.

- **Capabilities**

Capabilities traditionally have been segmented into discrete sources of advantage, such as knowledge, skills, and processes. AI-driven automation merges these areas in a continual cycle of execution, exploration, and learning. As an algorithm incorporates more data, the quality of its output improves. Similarly, on the human side, agile ways of working blur distinctions between traditional capabilities as cross-functional teams build quick prototypes and improve them on the basis of fast feedback from customers and end users.

AI and agile are inherently iterative. In both, offerings and processes become continuous cycles. Algorithms learn from experience, allowing companies to merge the broad and fast exploration of new opportunities with the exploitation of known ones. This helps companies thrive under conditions of high uncertainty and rapid change.

In addition to reframing specific sources of competitive advantage, AI helps increase the rate and quality of decision making. For specific tasks, the number of inputs and the speed of processing for machines can be millions of times higher than they are for humans. Predictive analytics and objective data replace gut feel and experience as a central driver of many decisions. Stock trading, online advertising, and supply chain management and pricing in retail have all moved sharply in this direction.

To be clear, humans will not become obsolete, even if there will be dislocations similar to (but arguably more rapid than) those during the Industrial Revolution. First, you need people to build the systems. Uber, for instance, has hired hundreds of self-driving vehicle experts, about 50 of whom are from Carnegie Mellon University's Robotics Institute. And AI experts are the most in-demand hires on Wall Street. Second, humans can provide the common sense, social skills, and intuition that machines currently lack. Even if routine tasks are delegated to computers, people will stay in the loop for a long time to ensure quality.

In this new AI-inspired world, where the sources of advantage have been transformed, strategic issues morph into organizational, technological, and knowledge issues, and vice versa. Structural flexibility and agility—for both man and machine—become imperative to address the rate and degree of change.

Scalable hardware and adaptive software provide the foundation for AI systems to take advantage of scale and flexibility. One common approach is to build a central intelligence engine and decentralized semiautonomous agents. Tesla's self-driving cars, for example, feed data into a central unit that periodically updates the decentralized software.

Winning strategies put a premium on agility, flexible employment, and continual training and education. AI-focused companies rarely have an army of traditional employees on their payroll. Open innovation and contracting agreements proliferate. As the chief operating officer of an innovative mobile bank admitted, his biggest struggle was to transform members of his leadership team into skilled managers of both people and robots.

• Getting Started

Companies looking to achieve a competitive edge through AI need to work through the implications of machines that can learn, conduct human interactions, and engage in other high-level functions—at unmatched scale and speed. They need to identify what machines do better than humans and vice versa, develop complementary roles and responsibilities for each, and redesign processes accordingly. AI often requires, for example, a new structure, of both centralized and decentralized activities, that can be challenging to implement. Finally, companies need to embrace the adaptive and agile ways of working and setting strategy that are common at startups and AI pioneers. All companies might benefit from this approach, but it is mandatory for AI-enabled processes, which undergo constant learning and adaptation for both man and machine.

Executives need to identify where AI can create the most significant and durable advantage. At the highest level, AI is well suited to areas with huge amounts of data, such as retail, and to routine tasks, such as pricing. But that heuristic oversimplifies the playing field. Increasingly, all corporate activities are awash in data and capable of being broken down into simple tasks. (See Exhibit 3.) We advocate looking at AI through four lenses:

- Customer needs
- Technological advances
- Data sources
- Decomposition of processes

First, define the needs of your customers. AI may be a sexy field, but it always makes sense to return to the basics in building a business. Where do your current or potential customers have explicit or implicit unmet needs? Even the most disruptive recent business ideas, such as Uber and Airbnb, address people's fundamental requirements.

Second, incorporate technological advances. The most significant developments in AI generally involve assembling and processing new sources of data and making partially autonomous decisions. Numerous services and platforms can capture incoming data from databases, optical signals, text, and speech. You will probably not have to build such systems yourself. The same is true on the back end as a result of the increasing availability of output technologies such as digital agents and robots. Consider how you can use such technologies to transform your processes and offerings.

Third, create a holistic architecture that combines existing data with new or novel sources, even if they come from outside. The stack of AI services has become reasonably standardized and is increasingly accessible through intuitive tools. Even nonexperts can use large data sets.

Finally, break down processes and offerings into relatively routinized and isolated elements that can be automated, taking advantage of technological advances and data sources. Then, reassemble them to better meet your customers' needs.

For many organizations, these steps can be challenging. To apply the four lenses systematically, companies need to be familiar with the current and emerging capabilities of the technology and the required infrastructure. A center for excellence can serve as a place to incubate technical and business acumen and disseminate AI expertise throughout the organization. But ultimately, AI belongs in and belongs to the businesses and functions that must put it to use.

Only when humans and machines solve problems together—and learn from each other—can the full potential of AI be achieved.

The Future of AI, and its Implications

Galileo viewed nature as a book written in the language of mathematics and decipherable through physics. His metaphor may have been a stretch for his milieu, but not for ours. Ours is a world of digits that must be read through computer science.

It is a world in which artificial-intelligence (AI) applications perform many tasks better than we can. Like fish in water, digital technologies are our infosphere's true natives, while we analog organisms try to adapt to a new habitat, one that has come to include a mix of analog and digital components.

We are sharing the infosphere with artificial agents that are increasingly smart, autonomous, and even social. Some of these agents are already right in front of us, and others are discernible on the horizon, while later generations are unforeseeable. And the most profound implication of this epochal change may be that we are most likely only at the beginning of it.

The AI agents that have already arrived come in soft forms, such as apps, web bots, algorithms, and software of all kinds; and hard forms, such as robots, driverless cars, smart watches, and other gadgets. They are replacing even white-collar workers, and performing functions that, just a few years ago, were considered off-limits for technological disruption: cataloguing images, translating documents, interpreting radiographs, flying drones, extracting new information from huge data sets, and so forth.

Digital technologies and automation have been replacing workers in agriculture and manufacturing for decades; now they are coming to the services sector. More old jobs will continue to disappear, and while we can only guess at the scale of the coming disruption, we should assume that it will be profound. Any job in which people serve as an interface – between, say, a GPS and a car, documents in different languages, ingredients and a finished dish, or symptoms and a corresponding disease – is now at risk.

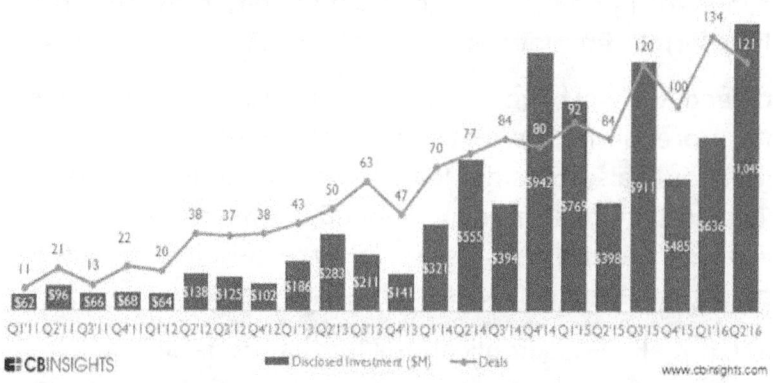

Image: CB Insights

But, at the same time, new jobs will appear, because we will need new interfaces between automated services, websites, AI applications, and so forth. Someone will need to ensure that the AI service's translations are accurate and reliable.

What's more, many tasks will not be cost-effective for AI applications. For example, Amazon's Mechanical Turk program claims to give its customers "access to more than 500,000 workers from 190 countries," and is marketed as a form of "artificial artificial intelligence." But as the repetition indicates, the human "Turks" are performing brainless tasks, and being paid pennies.

These workers are in no position to turn down a job. The risk is that AI will only continue to polarize our societies – between haves and never-will-haves – if we do not manage its effects. It is not hard to imagine a future social hierarchy that places a few patricians above both the machines and a massive new underclass of plebs. Meanwhile, as jobs go, so will tax revenues; and it is unlikely that the companies profiting from AI will willingly step in to support adequate social-welfare programs for their former employees.

Instead, we will have to do something to make companies pay more, perhaps with a "robo-tax" on AI applications. We should also consider legislation and regulations to keep certain jobs "human." Indeed, such measures are also why driverless trains are still rare, despite being more manageable than driverless taxis or buses.

Still, not all of AI's implications for the future are so obvious. Some old jobs will survive, even when a machine is doing most of the work: a gardener who delegates cutting the grass to a "smart" lawnmower will simply have more time to focus on other things, such as landscape design. At the same time, other tasks will be delegated back to us to perform (for free) as users, such as in the self-checkout lane at the supermarket.

Another source of uncertainty concerns the point at which AI is no longer controlled by a guild of technicians and managers. What will happen when AI becomes "democratized" and is available to billions of people on their smartphones or some other device?

What's more, many tasks will not be cost-effective for AI applications. For example, Amazon's Mechanical Turk program claims to give its customers "access to more than

500,000 workers from 190 countries," and is marketed as a form of "artificial artificial intelligence." But as the repetition indicates, the human "Turks" are performing brainless tasks, and being paid pennies.

These workers are in no position to turn down a job. The risk is that AI will only continue to polarize our societies – between haves and never-will-haves – if we do not manage its effects. It is not hard to imagine a future social hierarchy that places a few patricians above both the machines and a massive new underclass of plebs. Meanwhile, as jobs go, so will tax revenues; and it is unlikely that the companies profiting from AI will willingly step in to support adequate social-welfare programs for their former employees.

Instead, we will have to do something to make companies pay more, perhaps with a "robo-tax" on AI applications. We should also consider legislation and regulations to keep certain jobs "human." Indeed, such measures are also why driverless trains are still rare, despite being more manageable than driverless taxis or buses.

Still, not all of AI's implications for the future are so obvious. Some old jobs will survive, even when a machine is doing most of the work: a gardener who delegates cutting the grass to a "smart" lawnmower will simply have more time to focus on other things, such as landscape design. At the same time, other tasks will be delegated back to us to perform (for free) as users, such as in the self-checkout lane at the supermarket.

Another source of uncertainty concerns the point at which AI is no longer controlled by a guild of technicians and managers. What will happen when AI becomes "democratized" and is available to billions of people on their smartphones or some other device?

All of these profound transformations oblige us to reflect seriously on who we are, could be, and would like to become. AI will challenge the exalted status we have conferred on our species. While I do not think that we are wrong to consider ourselves exceptional, I suspect that AI will help us identify the irreproducible, strictly human elements of our existence, and make us realize that we are exceptional only insofar as we are successfully dysfunctional.

In the great software of the universe, we will remain a beautiful bug, and AI will increasingly become a normal feature.

www.ingramcontent.com/pod-product-compliance
Lightning Source LLC
Chambersburg PA
CBHW061437180526
45170CB00004B/1445